Reasoning About Theoretical Entities

Advances in Logic

Series Editor: Dov M Gabbay FRSC FAvH
Department of Computer Science
King's College London
Strand, London WC2R 2LS
UK
dg@dcs.kcl.ac.uk

Published

Vol. 1 Essays on Non-Classical Logic
by H. Wansing

Vol. 2 Fork Algebras in Algebra, Logic and Computer Science
by M. Frias

Vol. 3 Reasoning about Theoretical Entities
by T. Forster

Advances in Logic – Vol. 3

Reasoning About Theoretical Entities

Thomas Forster
University of Cambridge, UK

World Scientific
New Jersey • London • Singapore • Hong Kong

Published by

World Scientific Publishing Co. Pte. Ltd.

5 Toh Tuck Link, Singapore 596224

USA office: Suite 202, 1060 Main Street, River Edge, NJ 07661

UK office: 57 Shelton Street, Covent Garden, London WC2H 9HE

British Library Cataloguing-in-Publication Data
A catalogue record for this book is available from the British Library.

REASONING ABOUT THEORETICAL ENTITIES
Advances in Logic — Volume 3

ISBN 981-238-567-3

Printed in Singapore by Mainland Press

Preface

In this essay I am attempting to give a clear and comprehensive (and comprehensible!) exposition of the formal logic that underlies reductionist treatments of various topics in post-nineteenth-century analytic philosophy. The aim is to explain in detail—in a number of simple yet instructive cases—how it might happen that talk about some range of putative entities could be meaningful, have truth conditions and so on, even if those entities should be spurious. Although this ontological position has been adopted in relation to a wide range of putative entities at various times by various people I develop the logical gadgetry here quite specifically in connection with one such move: cardinal and ordinal numbers as virtual objects and always with the Burali-Forti paradox in mind.

Such a position (with respect to numbers at least) is one I associate with the work of Quine ("The subtle point is that any progression will serve as a version of number so long and only so long as we stick to one and the same progression. Arithmetic is, in this sense, all there is to number: there is no saying absolutely what the numbers are; there is only arithmetic.") though I think it is associated in the minds of many others with Dedekind. Indeed it seems to me to be wider than that, and to be an implicit part of the tradition. So implicit, and deemed perhaps to be so obvious, that nobody—as far as I know—has bothered to spell it out. This dereliction has had bad consequences. In my experience the gravest of these is that this treatment of numbers as virtual objects is confused with a much more widely (mis)understood position, namely the view associated with Frege that natural numbers are equivalence classes of sets under equipollence. I demarcate this from the position I explain in this book not because I believe it to be mistaken, but merely because it invites confusion with the position I am setting out to explain.

Thus we are committed to developing a logic of virtual objects. The particular case that I shall have permanently in mind in this essay is ordinal numbers and the Burali-Forti paradox. The Burali-Forti paradox, more than any of the others, requires for its resolution a proper treatment of virtual entities. It is only by treating (ordinal) numbers as virtual entities properly that we understand what the mistakes are that gave rise to that paradox. And mistakes I believe there have been, since I do not belong to the school of thought that thinks that paradox is part of the web of belief.

Thanks are due to the University of Genoa for inviting me to give the lectures in 1992 that were to become the start of this book and to Dr Giuseppe Rosolini for organising my trip. I am also grateful to the University of California at Irvine for three months sunsoaked leisure in their spring quarter in 2001 that I was able to devote to work on this book, and to Pen Maddy, Aldo Antonelli and Barbara Atwell for making it possible. It is a pleasure also to be able to thank the Royal Society for defraying the costs of a visit to Harvard to consult Professor Quine about these and other matters. I would like also to thank the Victoria University of Wellington, Massey University and the University of Auckland for inviting me to talk about this topic at their various seminars. Finally Paul Taylor earns my thanks—as he has earned the thanks of thousands of logicians over the years—for supplying the LaTeX macros that print commutative diagrams.

Thomas E. Forster
Department of Pure Mathematics
and Mathematical Sciences
University of Cambridge
Wilberforce Road, CB3 OWB
Cambridge, United Kingdom

Contents

Chapter 1

Introduction

Although I am using the Burali-Forti paradox as a case study, a kind of peg on which to hang observations about reductionism, this is not a history of the paradox, nor a picture of philosophy of mathematics as it was at the time the paradox emerged. Somebody ought to write such a book. Until that is done, readers who want that should consult Moore [1982]. This essay is a thorough deconstruction of the Burali-Forti paradox in a mainstream post nineteenth century logical tradition. It does not claim to be radical, but merely to make clear what the received view on this topic would be if the consensus of informed opinion had troubled to formulate itself. Why write such a book if it has no radical insights to offer? Because although the analysis is elementary, it is neither obvious nor particularly easy. If it were easy it would have been done long ago, and it were obvious there would be no need to do it at all.

When I went to university I started doing a degree in philosophy. In those days we were exposed to logical positivism: later generations brought up on different and more modern (or post-modern) isms may scoff, but for those interested in understanding how the world works it is probably as good a point of departure as any. It was then that I encountered the doctrine that physical objects are logical constructions out of sense data. It seemed painfully clear to me even at that early stage that if I really wished to even *understand* this thesis (let alone decide whether or not it was true) then I would have to learn about these *logical constructions*—whatever they might be—so I set about learning some logic.[1]

The doctrine that physical objects are logical constructions out of sense data is not at all atypical in modern philosophy. Claims with this flavour are made all the time: "meaning is use"; the sociobiology and sociobiology-

[1] I remember being particularly struck by the section on the theory of definitions in Suppes' Introduction to Logic, then hot from the press, now recently and justly reissued as a classic by Dover. [1957].

inspired literature is full of warnings that attributions of intention to genes are merely a *façon de parler*; the history of ethics is replete with claims that ethical judgements "just are" linguistically transformed assertions of some quite different and supposedly simpler kind, such as commands, or perhaps roars of approval or howls of disgust. According to (ethical) emotivism moral judgements are syntactic sugar for expressions of disgust or approval. There are cases that make it look plausible: my favourite example for years was the moral pronouncements made by homophobes, to the effect that gay sex is wicked and that gays should be punished. It certainly seems very plausible that the moral judgements expressed by homophobes are merely expressions of revulsion. So perhaps something analogous goes for moral judgements in general.

Plausible though this looks, I have come to the conclusion that it is wrong. Homophobes are indeed expressing disgust, but they do not thereby succeed in expressing a moral judgement *despite their use of moral language to do it*! What is going on is not (i) that they are using moral language appropriately and emotivism is true, but rather (ii) that they are using moral language in a nonstandard way, borrowing it to lend force to their utterances, and emotivism is false. Any discourse can be used metaphorically, and a discourse that has power riding on it is more likely to be used metaphorically than others! So the the superficially moral discourse of homophobes does not give us a fact about the content of moral language (which would be interesting) but merely a fact about the uses to which it can be put. And when you reflect that metaphor is inexhaustible, it hardly comes as a surprise to discover that the rhetoric of morality can be used to express homophobia. This is a fact about homophobes, not morality: it is not a fact about about moral language (that it has no propositional content) but a fact about the way that homophobes have of using it.

After all, if I use the language of widgets metaphorically to make a point about gadgets, this doesn't mean that widgets are just syntactic sugar for gadgets. One must distinguish between the sentence and its uses.

Even now that the legacy of the Vienna school has decayed there remains to us a feeling that we should try to think of each subject in the following list as being a part of its predecessor simplified with contextual definition: physics, chemistry, biochemistry, psychology, economics, history. This book is aimed at people who want to make a start on finding out what these claims mean. My aim in writing it has not been primarily to defend a reductive account of ordinals but to save others the bother of working out precisely what logic-chopping such a reduction involves. I am no reductionist: for me reductionism is a strategy for flushing out ontological commitment. I share with the anti-reductionists a hunch that reductionism won't work. What I do not share is their superstition that it is possible to understand the limitations of reductionist strategies without actually acquiring enough logic to formally execute them. This is an error: the belief that something won't work is not automatically a reason for not trying it, for even if failure is certain the manner of it might be instructive. In any case, as Dana Scott wrote [1962] "Nothing supports belief like proof". This error arises in part from a tendency to overestimate the importance of aquaintance with facts and underestimate the importance of *skills*. This is the error of thinking that all knowledge is *knowledge-that* not *knowledge-how*. It is an error with beguiling appeal, particularly to those who, for reasons of laziness or incapacity, are averse to the effort involved in concretising their understanding of reductionism, but it is an error nonetheless. The anti-reductionists differ from me not in our beliefs about reductionism, but in the lengths they are prepared to go to test them. Unwillingness to act on this hunch—together with the need for *post hoc* justifications of that unwillingness—call forth from the anti-reductionists a range of (*post hoc!*) *a priori* explanations for the supposed vanity of the enterprise (physicalist explanations cannot capture *qualia* etc., etc., ...), one and all spurious. The only way to understand what reductionist strategies can (and cannot) do is to learn how to execute them. This is not a fact about philosophy, but a fact about the human brain and how it acquires knowledge.

One effect of the tendency of some anti-reductionists to dismiss reductionism out of hand is a failure to explain properly what it is that is being dismissed, so that newcomers to the debate never really discover what is at issue. A recent textbook summarises reductionism as follows: "Reduction, as it is understood in most philosophical accounts involves deriving one theory from another, where theories are construed as sets of laws" (Bechtel *et al.* [2001]) (The use of the word 'construed' flags this text as a piece of philosophical discourse, but the expectations it raises are not met.)

This is not to say that there cannot be principled reasons for believing that reductionism is a vain enterprise. There may be. I suspect that the impulse to believe it is doomed stems from a cause whose best illustration is the old saw that one cannot derive an 'ought' from an 'is'. This is indeed starkly obvious, and to those of us brought up in twentieth-century logic the obviousness seems to arise from the interpolation lemma. This states that if we have derived a formula Q from a formula P, then there is a formula R. derivable from P. and from which Q in turn can be derived, which contains only vocabulary common to P and Q. The simplest case of the interpolation lemma concerns the case where $P \to Q$ is a tautology but P and Q have no nonlogical vocabulary in common. In this case either Q is a tautology or P is the negation of one. And this case is the one Hume had in mind: after all, the vocabularies of physics and of moralising are disjoint, and so only in trivial cases can one deduce an 'ought' form an 'is'. The interpolation lemma is one of the most intuitively obvious results of modern logic, and the intuition which makes it obvious is at work even in people who know no logic, and in them it results in a conviction that reductionist strategies must fail. And yet, one wants to say, it is true both that everything has to emerge from 'is's and that we can't just write morality off. One way round this is to attempt to interpret moral language in physicalist language. One cannot *derive* an 'ought' from an 'is' but one might be able to *interpret* an 'ought' as an 'is'.[2]

The conditions controlling use of a contextually defined term can take the form of syntactic constraints. These syntactic constraints can in turn take the form of type-theoretic distinctions of the kind that were first explicitly developed by Russell [1905]. Since that time a type-theoretic hierarchy of some kind has traditionally been invoked as a way of dealing with the paradoxes that so plagued the foundations of mathematics at that time. It does not nowadays take the genius of a Russell to see that this was a good idea. Indeed in contrast nowadays (and especially since Kripke [1976]) we are so *blasé* about this insight that it now almost seems to be a good idea whose time has gone: it has become fashionable among philosophers to decry type-theoretic critiques of the paradoxes. Although an approach that eschews type-theory looks quite plausible for the semantic paradoxes (which are those that philosophers are more likely to be concerned with) it does not work nearly so well with the logical paradoxes (and the Burali-Forti paradox in particular), which have much deeper roots in hierarchy theory. It is not hard to see how this fact came to be overlooked, for the Burali-Forti paradox is by far the most obscure of the paradoxes: even Burali-Forti, when writing the article (Burali-Forti [1897]) in virtue of which the paradox bears his name, did not think that he had discovered a proof of a contradiction, but a *reductio ad absurdum* proof that ordinals are not totally ordered. However—and this important point will become clearer as this essay proceeds—the type-theoretical hierachy that we need to exploit in order to avoid the Burali-Forti paradox has its roots in the syntactic disciplines entailed by contextual definitions, and has nothing to do with the type distinctions of Russell [1905].

The logical positivists used the expression 'category mistake' to capture some of these oddities. My favourite example is Carnap's illustration: "This stone is thinking about Vienna" (which reveals where the example comes from!). "X is thinking about Y" is (or can be analysed as) an abbreviation

[2]Parallels can be found in the literature in hardware verification. Correctness assertions seem to belong to a different vocabulary from that used to describe circuitry, and the interpolation lemma makes itself felt again. the effect that correctness proofs can be of no effect. And again, the response is to look for ways of interpreting the second vocabulary in the first. See Melham chapter 4, and Eveking, [1991]

for a long conjunction of assertions about X. The point is that the linguistic rules will allow this abbreviation to be introduced only if certain things are true, and the vast majority of those things are false of stones. I free-associate from this quotation to memories of discussions of the truth-value status (True? False? Meaningless?) of such monsters. I shall not be joining such a debate here: my purpose is to show how the Burali-Forti paradox can be seen as arising from a failure to respect the conditions attached to a contextual definition in a similar way, and partakes of the absurdity—not just of the other paradoxes—but also of the assertion that a stone is thinking about Vienna. In some ways this is a more instructive pastime: although the conventions surrounding the use of ordinal notations can be made explicit in a way that the conventions breached by talk of a stone contemplating Vienna cannot (for the moment!) they have not so far been made thus explicit.

There is another cause of people overlooking the deeper roots that Burali-Forti has in type-theory. For many years I for one believed that type-theory was primarily a way of avoiding the paradoxes. It was not until I started teaching computer science and was forced to think about a distinction that for want of a better word one might call an *essence-implementation* distinction that I came to realise that the purpose of respecting type disciplines is to avoid making fools of ourselves by overlooking it. If we respect type disciplines in our use of language, then we can avoid asking silly questions like: "Is there an x identical with the ordered pair $\langle x, x\rangle$?" Although there is a risk of self-centredness in assuming that others will have made the same mistakes one has made oneself, one can be very useful to them if one happens to be right! It now seems to me that type-theory plays a much more central rôle in the dissolution of the Burali-Forti paradox than it does in any of the others. Indeed I would claim that it is only a type-theoretic critique of the Burali-Forti paradox that enables us to understand it at all. By "type-theoretic" I do not mean 'type-theory' as in what the modern radicals ridicule as the "Tarski hierarchy" but 'type-theory' as in the syntactic disciplines we observe in order to avoid impaling ourselves on questions like "$x = \langle x, x\rangle$?", namely "type-theory" as understood by people in theoretical computer science.

The conclusion of this line of thought is that the modern heresy that type-theoretic approaches to the paradoxes are wrong (and incoherent) is not likely to be fruitful. If there were any possibility of freeing ourselves from type-theory altogether there would be some point in the thought-experiment of approaching the paradoxes armed instead with some other weapon. But since we are always going to have to respect type distinctions of some kind—if for no other reason than that type-distinctions are forced on us by contextual definitions (and it is *those* type distinctions that are at the root of the Burali-Forti paradox)—we may as well try to put them to good use.

To understand the rôle of contextual definitions in the manufacture of theoretical entities it is an idea to start with a case we understand. First-order cardinal arithmetic is probably the most complicated instance of this phenomenon for which there exists a community of scholars who understand—in outline—its status as a theory of contextually defined entities. Higher-order ordinal arithmetic is probably the most complicated instance of this phenomenon for which it is possible to give an exhaustive account at this stage. Instances of wider philosophical interest are unfortunately too complicated to treat properly for the time being. The reason for placing the Burali-Forti paradox in a general philosophical context in this (what might be felt to be tendentious) way is that although the Burali-Forti paradox is of considerable independent interest it can also be thought of as a toy version of the puzzle of contextually defined entities which permeates analytical philosophy.

However there is one respect in which it is of greater interest. There are several suites of entities admitting reductive analyses arsing from congruence relations which are generally well understood. One thinks of the various standard presentations of integers over naturals, rationals over integers, reals over rationals, vectors as equivalence classes of directed line segments (see Lederman-Vajda *op. cit.* for example). In all these cases the equivalence classes—indeed even the quotients themselves, the sets of equivalence classes—are sets according to even the most penny-pinching comprehension axioms, so one never has to concern oneself with the difference between simulations and implementations which will be important below (and which will inevitably be important in other context such as philosophy of mind where one might want a reductive story that really does say that certain things do not exist). The other difference is that any reduction that gives us cardinals, will also give us sets of cardinals. The result of this is that there will be cardinals of sets of cardinals: the reduction can be iterated. As we will see below, this has important ramifications.

Summary

We start with a quick survey of interpretations, rapidly restrict our attention to interpretations that arise from congruence relations, (but not before we have had a brief look at the manner in which definite descriptions can be added to a theory obtain a new theory interpretable in the old) and see how these can give rise to first- (and higher-) order theories of the original entities with new virtual entities. A distinction is made between interpretations that respect equality and those that do not. Two examples (cardinal and ordinal arithmetic) are pursued in some detail. Some time is spent on sketching which parts of what is commonly regarded as cardinal and ordinal arithmetic can be interpreted back into the set theory from which they arise in a manner that does not depend on the ways in which cardinals and ordinals are implemented. This leads to some discussion of set-existence issues which, although peripheral to the rhetorical purpose of the essay, are nevertheless of some interest to students of philosophy of mathematics.

Finally a word or two is in order on what is *not* being done. Although the endeavour is to shed some light on how assertions apparently about things that do not exist can have truth-conditions and a sensible semantics, this treatment is not intended to explain how statements about fictional figures can have genuine truth-conditions (or not, as instance L.C.Knights famous question "How many children had Lady Macbeth?") nor is it intended to cover truths about nonexistent objects of the kind considered by Condoravadi *et al.*

Historical remarks

There is no doubt that the Burali-Forti paradox is the most obscure of the classical paradoxes. Indeed, at its first appearance it was not even recognised as a paradox, but believed to be a *reductio ad absurdum* proof that ordinal inequality is not trichotomous. (This is the negation of theorem 6.4.) Russell's view in 1901-3 when he was writing his *Principles of Mathematics* [1903] (see ch XXXVIII paragraph 301) was that it proved that the order relation on ordinals is not wellfounded. (This is theorem 6.4.) By the time he was writing *Principia Mathematica* with Whitehead, his view had changed, and he had come to the modern view. (See volume III of *Principia Mathematica* pp 74-5). Even as late as 1940 it was possible for as distinguished a scholar as Quine to publish an axiomatisation of set theory in which an implementation of the paradox could be proved. (See p. 85.) The fact that the standard proofs of all the other paradoxes had been successfully blocked underlines how obscure the Burali-Forti paradox is. Russell (Introduction to Mathematical Philosophy) doesn't mention it at all.

Notation

We use standard set-theoretic notation here. $\emptyset$ is the empty set, $\bigcup x$ is $\{y : (\exists z)(y \in z \wedge z \in x\}$, $\bigcap x$ is $\{y : (\forall z)(z \in x \rightarrow y \in z\}$, $f(x)$ is the value the function f gives to the argument x, to contrast it with $f``x$ which is the set of values the function f gives to the set of arguments x. We will sometimes omit the brackets from functional application where a λ-calculus tradition (which eschews them) seems more stylish. Thus '$\texttt{fst}\langle x, y\rangle$ ' rather than '$\texttt{fst}(\langle x, y\rangle)$'. $\lambda x.[...]$ is the function that, on being given x, returns the stuff enclosed by the square brackets. A superscripted arrow on a variable means a list of variables, thus: $\vec{x}$. '■' signals the end of a proof. Following Russell, we use upside-down iotas for singular descriptions. Expressions written in **bold** are being defined; expression in *italic* and being emphasised.

Quine quotes ("corners") are a device for obtaining variable names for expressions with variables in them. They should be thought of as a diacritic which creates a context within which names of expressions can be combined with connectives to give names of the expressions thus combined.

We will use upper case Roman letters for variables ('X', 'X'', 'Y' ...) ranging over sets. Lower case variables ('x', 'y'', 'y' ...) will range over whatever it is that those sets are sets of. We will have these two styles—common in set theory—so that we can write '$x \in X$' as usual. Capitalised *fraktur* font variables will range over **structures**. If $\mathfrak{X}$ is a mathematical structure consisting of a set X (the **domain**) and a relation R on X we will say that $X = DM(\mathfrak{X})$ and $R = DG(\mathfrak{X})$, and the letter denoting the domain of a structure will be the uppercase Roman letter corresponding to the uppercase fraktur letter denoting the structure. A **reduction** of a structure is a structure with the same domain but some operations or predicates removed. An **expansion** of a structure is a structure with the same domain and operations with some extra ones added. See (e.g.) Hodges [1993]

1.1 Interpretations

An interpretation $\mathcal{I}$ of a theory T in a theory S is a recursive map from $\mathcal{L}_T$ to $\mathcal{L}_S$ which sends elements of T to elements of S. By 'recursive' I mean that it should be defined by recursion on the grammar of $\mathcal{L}_T$ rather than that, considered as a map from Gödel numbers to Gödel numbers, it should have a decidable graph. Announcing this restriction shouldn't start any fights: there may be circumstances in which nonrecursive maps need to be taken into account, but they destroy information too comprehensively to be any help in an analysis intended to allow understanding to flow through the interpretation from the interpreting language to the interpreted. In fact the nature of the interpretations we consider will be restricted still further.

Clearly $\mathcal{I}$ must send atomic n-place relation (resp. function) symbols of $\mathcal{L}_T$ to (possibly derived) n-place relation (resp. function) symbols of $\mathcal{L}_S$. Typically it will be the case that $\mathcal{L}_T$ is a superset of $\mathcal{L}_S$. For example $\mathcal{L}_S$ might be the language of set theory, and $\mathcal{L}_T$ the language of set theory plus arithmetic. Again $\mathcal{L}_S$ might be the language of physics, and $\mathcal{L}_T$ the language of physics plus some mentalistic language.

$\mathcal{I}$ might satisfy other sensible conditions, for example it might send $=$ to $=$. The condition that $\mathcal{I}$ be recursive suggests (though it does not entail) that $\mathcal{I}$ should commute with quantifiers and connectives. (If $\mathcal{I}$ is recursive we know that $\mathcal{I}$ of $\Psi \wedge \Phi$ must be $(\mathcal{I}(\Psi))??(\mathcal{I}(\Phi))$ for some binary connective ??.) One example of an interpretation that commutes in this way is the interpretation of meta-arithmetic into arithmetic that is needed for the proof of the incompleteness theorem. A family of examples of interpretations that do not so commute are the negative interpretations of classical logic into intuitionistic logic. Others include the Gödel interpretation of a modal propositional language in a nonmodal one and the interpretation of a classical dyadic language into an intuitionistic monadic one.

Whether or not an interpretation sends atomics to atomics or commutes with logical operators is not going to matter a great deal here. Although we will distinguish two kinds of interpretations we will distinguish them on different grounds: those I shall call *implementations* and those I shall call *simulations*.

An interpretation of $\mathcal{L}_1$ into $\mathcal{L}_2$ that sends equality to equality is an **implementation**. All other interpretations are **simulations**.

When we have a simulation of a language $\mathcal{L}_1$ into another language $\mathcal{L}_2$ we shall say that from the point of view of $\mathcal{L}_2$ the entities of $\mathcal{L}_1$ are **virtual**.

In the situations that concern us here we will be interested in simulations where the interpretation of equality is a equivalence relation which is natural in that context. For example we will simulate cardinal arithmetic in set theory by interpreting equality between cardinals as equinumerosity (aka equipollence) between sets. Such simulations we will call *canonical*.

The situations in which one looks for interpretations fall naturally into one of two classes, depending on whether on not the entities one wishes to implement are to be "magicked away" or not. A familiar example of this is the interpretation of elementary-arithmetic-with-pairing-and-unpairing into elementary arithmetic. One can implement $\langle x, y\rangle$ as $2^x \times (2y+1)$ or $\binom{x+y+1}{2} + x$ or in any of a host of other ways. In set theory it is customary to interpret set-theory-with-pairing-and-unpairing inside set theory by the Wiener-Kuratowski device of taking $\langle x, y\rangle$ to be $\{\{x\}, \{x, y\}\}$, though it is reasonably well known that there are other ways of doing it. Russell famously showed how to interpret languages with singular descriptions into languages without singular descriptions. In neither of these cases is there any suggestion that ordered pairs are not real, or that the things singularly identified are not real. In these cases the interpretation involved is an implementation. This is in contrast to cases (like those that will occupy most of this book) where the language being interpreted speaks of entities that are not present in the language into which the interpretation is directed, and the interpretation is a *simulation*. Those are the cases where there seems to be a genuine possibility of ontological reduction. "There are no numbers, there is only arithmetic." (to misquote Quine).

In circumstances like this it is natural when considering expressions in the language of (for example) set-theory-with-pairing-and-unpairing to ascribe a privileged position to those expressions ϕ such that the truth-value of the implementation of ϕ into set theory does not depend in any way on the implementation (of pairing and unpairing) chosen. It's fairly plausible that these expressions can be given a purely syntactical characterisation. The expressions that fail the syntactic test (like $(\exists x)(x = \langle x, 0\rangle)$) also have the flavour of puns, but this time for a subtly different reason. The point is not that the canonical simulation is not defined on them (true though that is, since there is no canonical simulation—indeed no simulation at all) but that their truth-value is implementation-dependent. Punning of this kind is an important and idiomatic part of some programming languages (specifically and importantly the languages C and C^{++}) and is the most important single cause of the distaste with which these languages are regarded by the *cognoscenti*. It also plays a central rôle in the proofs of the major results of recursion theory, most obviously the proof of the unsolvability of the halting problem, which relies on a number being both (i) treated as data and (ii) executed as the index of a machine. Much of what there is to say about the Gödel sentence does not depend on a choice of Gödel numbering. This is less true for Rosser sentences, and is one of the reasons why Rosser sentences are less well understood.

This semantical characterisation can be made also in cases—like that where $\mathcal{L}_1$ is the language of set theory and cardinals—where there are genuine interpretations in addition to a canonical simulation. (There is obviously no canonical simulation in the case of ordered pairs!) Will it turn out that the subset of the language of set theory and cardinals on which the canonical simulation is defined is the same as the subset which is implementation-independent? Irritatingly not, though the first is at least a subset of the second. We shall see examples of expressions that are implementation-independent but still not in the domain of the canonical simulation. An important example is the combinatorial assertion about $\mathbb{N}$ that lies at the heart of the Paris-Harrington theorem. Other examples pave the path to the Burali-Forti paradox.

Outside the strictly mathematical context which will be the main concern here we can see other areas where this simulation/implementation terminology will be useful. Part of the modern functionalist rescension of philosophy of mind is a view that mentalistic language is always an option which is available for describing in neat shorthand the gross behaviour of systems that exhibit (sufficiently complicated) feedback. For example Dennett (in [1979]) writes of "adopting the intensional stance". In the slang of computer scientists, mentalistic language is *syntactic sugar* for physicalist talk. In the terminology of the preceding paragraph, we are working in an (as it might be) physicalist language $\mathcal{L}_2$ and trying to interpret in it a mentalistic+physicalistic language $\mathcal{L}_1$. We *simulate* mental entities. What the various flavours of identity theory currently available in the philosophical literature haggle over is the nature of the *physical implementations* of the mental entities that the intensional language purports to describe.

There is a large class of interpretations which have philosophical interest. As we have noted, there are many claims of the kind "widgets are just gadgets", and in the terminology that prevails in this book claims like that would come out as "The theory of widgets and gadgets can be interpreted in the theory of gadgets". However not every claim of this kind is felt to have ontological repercussions in the way that most of us would feel that a reduction of sets-with-cardinals to sets shows that there aren't really any numbers at all. Certainly in the simplest case of all—singular descriptions, with which we begin the next chapter—there is no suggestion that the King of France fails to be real merely because the expressions that denote him and him alone can be magicked out of the language: the President of France can be magicked out of the language too, but he is real enough.

There are many virtual objects that one might be interested in studying and for which a robust and taut logical analysis is both needed and available. One thinks of a virtual theory of waves interpreted in a physics where the only real things are addresses in space-time and elementary particles.[3]

[3]This would be a classical theory not a quantum theory! In quantum theory waves are not virtual with respect to particles: they are equally real.

Chapter 2

Definite Descriptions

Although in most of this book we will be preoccupied with interpretations that arise from congruence relations, there is a very simple and very important example of an interpretation which we will need to master first, and this is Russell's treatment of singular descriptions. And we need to do this not merely because it is simple, or historically prior to other such analyses—though it is—but because we need to understand singular descriptions in order to give a smooth treatment of functions defined on virtual entities. Readers who are completely confident about Russell's theory of descriptions can probably safely skip the rest of this chapter.

In Russell's treatment of definite descriptions we replace expressions matching the template

$$(\exists x)(\psi(x) \wedge (\forall y)(\psi(y) \rightarrow y = x) \wedge \phi(x))$$

with expressions matching

$$\phi((\iota x)(\psi(x)))$$

(The vagueness in this description is deliberate: see below). As happens so often in Philosophical Logic, the analysis he gave can be used in two ways. On the one hand it can be used to develop formalised languages so that they exhibit features which behave in various helpful ways like features of natural languages (*i.e.*, in this case it explains how to introduce definite descriptions into formalised languages). On the other hand it can be used to make a claim about what the logical status of those natural language features *really* is. This distinction is sometimes called the *prescriptive-descriptive* distinction. Philosophers often take Russell's analysis in the second sense, thereby regarding it as a (descriptive) thesis about ordinary language to which reasonable people might take exception. (Strawson [1950] is an example.) Our primary concern here is with formalised languages (since part of the aim of the project is to illuminate the Burali-Forti paradox) so this second point of view will be touched only briefly.

In addition to the two uses to which Russell's analysis can be put, there are also two views one can have about what it actually says. The same two possibilities exist in relation to the indefinite descriptions of Hilbert, and there the possibilities of confusion are much more serious. It therefore seems a good idea to deal with the matter when it first comes up, when the context is simple.

On the one hand we can think of ι-terms as being introduced by contextual definition, so that $\phi((\iota x)(\psi(x))$ is short for $(\exists x)(\phi(x) \wedge \psi(x) \wedge (\forall y)(\psi(y) \to y = x))$. According to this version of events, the introduction of ι terms is an entirely orthographical move, and is nothing to do with logic at all. If one wants to think of Russell's account of ι-terms as a thesis about the logical structure of ordinary language, this compels one to say that the (top level) logical structure of "the King of France is bald" is an existential quantification. According to this view, ι-terms terms are syntactic sugar and do not really denote. However, we are not obliged to adhere to this purist view: in any situation in which an ι-term can legitimately be introduced there is a *canonical* implementation for it. This is in radical contrast to other contextually defined terms, as we shall see later.

On the other hand we can think of ι-terms as constants in a new language, belonging to a new theory which is a conservative extension of the old ("*no new theorems in the old vocabulary*"). The completeness theorem for first-order logic tells us that if we have a theory T expressed in a first-order language, and T proves an expression of the form $(\exists x)\psi(x)$ the extension T' of T obtained by adding a new constant to the language 'a', and adding an axiom $\psi(a)$ to T is conservative. *Therefore the coherence of this view of singular descriptions for a language $\mathcal{L}$ depends on the completeness theorem for $\mathcal{L}$.* Applying this version of Russell's analysis prescriptively to the examination of ordinary language we find that it tells us that "the King of France is bald" *really* has the structure of a predicate being attributed to a subject, *not* that of a existential quantification.

The precise status of claims like these is obscure: no-one can seriously pretend that these abbreviations are explicitly set up by a properly constituted parliament of language users. One rapidly finds oneself facing the same problems as social contract theorists do in ethics.

Although Strawson's critique is based on the assumption that Russell's reduction of singular descriptions is to be taken as an account of singular descriptions in ordinary language and therefore is not directly relevant here, this point of view has a formal correlate. The idea that "The King of France is bald" lacks truth-value when there is no unique king of France corresponds to a determination not to allow expressions containing $(\iota x)(\phi(x))$

to be well-formed when there is not precisely one thing that is ϕ. Normally in mathematical logic one sets up a language and then—on the understanding that this language is fixed—one adumbrates theories in it. In principle there is nothing to stop one having an arrangement in which the language is regarded as a dynamic entity, so that the language belonging to a theory is determined in part by that theory. In this case one would allow that if $T \vdash \exists! x\phi(x)$ then a constant '$\iota x.\phi(x)$' can be added to the language we started with. This may or may not have the merits of psychological plausibility and exciting technical complexity but it is not a path I intend to go down here. The obvious question of what one is to make of formulæ containing embedded definite descriptions whose introduction has not been authorised by a theorem in the style $\exists! x\phi(x)$ is largely sidestepped here by theorem 2.1.

In general $(\exists! x)(\exists! y)\phi(x, y)$ is not the same as $(\exists! y)(\exists! x)\phi(x, y)$: '$(\exists! x)(\exists! y)(y < x)$' is true in the natural numbers (the witnesses being 1 and 0) but '$(\exists! y)(\exists! x)(y < x)$' is not.

2.1 Formal definition of the interpretation

Suppose we have a language $\mathcal{L}$ which does not contain singular descriptions. We want to show how to interpret in $\mathcal{L}$ the language $\mathcal{L}^{\iota}$, the result of adding singular descriptions to $\mathcal{L}$.

Definition 2.1 Let $\mathcal{I}$ be the interpretation $\mathcal{L}^{\iota} \to \mathcal{L}$ defined as follows. The recursive clauses are that $\mathcal{I}$ commutes with quantifiers and connectives. Atomic and negatomic formulæ are sent to themselves unless they contain singular descriptions. Let us consider this case in more detail.

If 'x' is free in ϕ then for atomic or negatomic Ψ with 'y' free, $\mathcal{I}$ of $\Psi[\iota x.\phi(x)/y]$ is

$$(\exists x)(\phi(x) \wedge \Psi[x/y] \wedge (\forall y)(\phi(y) \to y = x))$$

If Ψ is an equation this last expression becomes

$$\phi(y) \wedge (\forall z)(\phi(z) \to z = y)$$

It is important that the rewriting rules we are implicitly defining by this declaration should be confluent! The expression $\phi((\iota x)A(x), (\iota y)B(y))$ contains two ι terms. We can think of this as $\Psi((\iota x)A(x))$ which we can then expand, getting an expression which contains an occurrence of $(\iota y)B(y)$, which we can then expand. On the other hand we could think of this as $\Psi((\iota y)B(y))$ which we can then expand, getting an expression which contains an occurrence of $(\iota x)A(x)$, which we can then expand.

It doesn't make any difference which order these two expansions are done in.

If we treat '$\phi((\iota x)A(x),(\iota y)B(y))$' as an assertion about $(\iota x)A(x)$ we expand it as follows. First to

$$(\exists x)(A(x) \wedge (\forall w)(A(w) \rightarrow w = x) \wedge \phi(x,(\iota y)B(y)))$$

Now expand $(\iota y)B(y)$ to get

$$(\exists x)(A(x) \wedge (\forall w)(A(w) \rightarrow w = x) \wedge (\exists y)(B(y) \wedge (\forall z)(B(z) \rightarrow z = y) \wedge \phi(x,y)))$$

and we can pull the '$\exists y$' to the front to get

$$(\exists x)(\exists y)(A(x) \wedge (\forall w)(A(w) \rightarrow w = x) \wedge B(y) \wedge (\forall z)(B(z) \rightarrow z = y) \wedge \phi(x,y))$$

We can always permute like quantifiers, so we do this, to get

$$(\exists y)(\exists x)(A(x) \wedge (\forall w)(A(w) \rightarrow w = x) \wedge B(y) \wedge (\forall z)(B(z) \rightarrow z = y) \wedge \phi(x,y))$$

...rearrange the matrix

$$(\exists y)(\exists x)(B(y) \wedge (\forall z)(B(z) \rightarrow z = y) \wedge A(x) \wedge (\forall w)(A(w) \rightarrow w = x) \wedge \phi(x,y))$$

...import the '$\exists x$'

$$(\exists y)(B(y) \wedge (\forall z)(B(z) \rightarrow z = y) \wedge (\exists x)(A(x) \wedge (\forall w)(A(w) \rightarrow w = x) \wedge \phi(x,y)))$$

abbreviate the third conjunct

$$(\exists y)(B(y) \wedge (\forall z)(B(z) \rightarrow z = y) \wedge \phi((\iota x)(A(x)),y)))$$

as desired.

We can now prove by induction on $\mathcal{L}$ that:

Theorem 2.1 *Suppose ϕ is built up from atomics and negatomics by means of $\wedge$, $\vee$, $\forall$ and $\exists$. Suppose also that 'x' is free in ϕ and no variable free in ϕ is bound by any quantifier in Ψ. Then for any Ψ with 'y' free, $\mathcal{I}$ of $\Psi[\iota x.\phi(x)/y]$ is logically equivalent to*

$$(\exists x)(\phi(x) \wedge \Psi[x/y] \wedge (\forall y)(\phi(y) \rightarrow y = x))$$

Proof:

This is proved by induction on Ψ. The base case (Ψ atomic or negatomic) follows immediately from definition 2.1. The inductive steps for the quantifiers and connectives are as follows.

Ψ is $\Psi_1 \vee \Psi_2$

Suppose 'x' is free in ϕ and no variable free in ϕ is bound by any quantifier in Ψ_1 or Ψ_2. By induction hypothesis $\Psi_1[\iota x.\phi(x)/y]$ is equivalent to $(\exists x)(\phi(x) \wedge \Psi_1[x/y] \wedge (\forall y)(\phi(y) \rightarrow y = x))$ and $\Psi_2[\iota x.\phi(x)/y]$ is equivalent to $(\exists x)(\phi(x) \wedge \Psi_2[x/y] \wedge (\forall y)(\phi(y) \rightarrow y = x))$. Therefore $(\Psi_1 \vee \Psi_2)[\iota x.\phi(x)/y]$ is equivalent to

$$(\exists x_1)(\phi(x_1) \wedge \Psi_1[x_1/y] \wedge (\forall y)(\phi(y) \rightarrow y = x_1)) \vee$$
$$(\exists x_2)(\phi(x_2) \wedge \Psi_2[x_2/y] \wedge (\forall y)(\phi(y) \rightarrow y = x_2))$$

By the uniqueness condition any witness to '$\exists x_1$' must also satisfy the first and third clauses of the second disjunct so the formula simplifies to

$$(\exists x)(\phi(x) \wedge (\Psi_1 \vee \Psi_2)[x/y] \wedge (\forall y)(\phi(y) \rightarrow y = x))$$

as desired

Ψ is $\Psi_1 \wedge \Psi_2$

Suppose 'x' is free in ϕ and no variable free in ϕ is bound by any quantifier in Ψ_1 or Ψ_2. By induction hypothesis $\Psi_1[\iota x.\phi(x)/y]$ is equivalent to $(\exists x)(\phi(x) \wedge \Psi_1[x/y] \wedge (\forall y)(\phi(y) \rightarrow y = x))$ and $\Psi_2[\iota x.\phi(x)/y]$ is equivalent to $(\exists x)(\phi(x) \wedge \Psi_2[x/y] \wedge (\forall y)(\phi(y) \rightarrow y = x))$. Therefore $(\Psi_1 \wedge \Psi_2)[\iota x.\phi(x)/y]$ is equivalent to

$$(\exists x)(\phi(x) \wedge \Psi_1[x/y] \wedge (\forall y)(\phi(y) \rightarrow y = x)) \wedge (\exists x)(\phi(x) \wedge$$
$$\Psi_2[x/y] \wedge (\forall y)(\phi(y) \rightarrow y = x)).$$

We want this to be equivalent to $(\exists x)(\phi(x) \wedge (\Psi_1 \wedge \Psi_2)[x/y] \wedge (\forall y)(\phi(y) \rightarrow y = x))$. Clearly this last formula implies both conjuncts. Conversely, if both conjuncts are true, the two existential quantifiers have the same witness, in which case this witness must have both Ψ_1 *and* Ψ_2.

The universal quantifier case

Suppose 'x' is free in ϕ and no variable free in ϕ is bound by any quantifier in Ψ but that 'z' is free in Ψ. By induction hypothesis $\Psi[\iota x.\phi(x)/y]$ is equivalent to $(\exists x)(\phi(x) \wedge \Psi[x/y] \wedge (\forall y)(\phi(y) \rightarrow y = x))$. So $(\forall z)\Psi[\iota x.\phi(x)/y]$ is equivalent to $(\forall z)(\exists x)(\phi(x) \wedge \Psi[x/y] \wedge (\forall y)(\phi(y) \rightarrow y = x))$. What we want next is that this should be equivalent to $(\exists x)(\forall z)(\phi(x) \wedge \Psi[x/y] \wedge (\forall y)(\phi(y) \rightarrow y = x))$

Normally we cannot pull existential quantifiers out past universal quantifiers in this way because the witness to the existential quantifier might depend on the choice of an instance for the universal quantifier. However in this case the uniqueness condition ensures that there is only one possible witness, and this enables us to pull the existential quantifier past the universal quantifier. Since 'z' is not free in '$\phi(x)$' or $(\forall y)(\phi(y) \rightarrow y = x)$ we can push the '$\forall z$' further inside to get $(\exists x)(\phi(x) \wedge (\forall z)\Psi[x/y] \wedge (\forall y)(\phi(y) \rightarrow y = x))$

The case with the existential quantifier is similar but easier.

■

For what it's worth this proof is constructive. Notice that this induction will not work for $\neg$ and $\rightarrow$. Formulæ built up from these connectives are not covered by this result. However every formula is classically equivalent to one that is. We need the clause that no variable free in ϕ is bound by any quantifier in Ψ for the following reason. If we take $\phi(x)$ to be "x is the mother of z" and Ψ to be '$(\forall z)(y$ is related to $z)$' then the result of applying the above rule to $\Psi[\iota x.\phi(x)/y]$ would be

> $(\exists x)(x$ is the mother of $z \wedge (\forall z)(x$ is related to $z) \wedge (\forall y)(y$ is the mother of $z \rightarrow y = x))$

which is clearly not what we want.

2.2 Functions from singular descriptions

Once we have an apparatus for introducing singular descriptions we can use this to introduce notations for functions. If $(\forall \vec{x})(\exists y)(\psi(\vec{x}, y) \wedge (\forall y')(\psi(\vec{x}, y') \rightarrow y = y'))$ then clearly we are justified in introducing the notation for the function connoted by the letter 'ψ'. Notice that nothing in theorem 2.1 depends on ϕ containing no free variables. Thus 2.1 applies to function letters as well as to singular descriptions, since **a (molecular) function symbol is simply a definite description with free variables.**

Henceforth we will assume without further comment that function letters can be introduced in this way. We will make use of it often in what is to follow, but always in circumstances where the value of the "function" is not unique but is unique-up-to-equivalence under the relevant equivalence relation.

In the long run it is important to understand the correct treatment of singular descriptions with parameters because that is one way of avoiding functional notation. In the short run, it enables us to make sense of an old puzzle of Quine's to which we now digress.

2.3 Definite descriptions and modal realism

Quine's puzzle is as follows. Consider the two terms '9' and 'the number of planets'. One might suppose they denote the same thing, but if they do, then anything true of (the denotation of) one is true of (the denotation of) the other. Unfortunately the first is necessarily greater than 7 and the second isn't. So either they aren't the same thing, or substitutivity of identicals fails.

To resolve this, we will have to ascertain the logical structure of "The number of planets is necessarily greater than 7". What is its main connective? Is it a box? Or is it a predication. If it is a predication we can unravel as follows:

A: $\Box((\iota x)(\text{number-of-planets}(x)) > 7)$

appears to match the pattern $\Psi((\iota x)(\phi(x)))$, (using the template of theorem 2.1) by matching 'number-of-planets(x)' to '$\phi(x)$' and '$\Box(x > 7)$' to '$\Psi(x)$'. We know what to do with this. It is just:

$$(\exists x) \bigwedge \begin{cases} \phi(x) \\ (\forall y)(\phi(y) \to y = x) \\ \Psi(x) \end{cases}$$

so A must be
A′:

$$(\exists x) \bigwedge \begin{cases} \text{number-of-planets}(x) \\ (\forall y)(\text{number-of-planets}(y) \to y = x) \\ \Box(x > 7) \end{cases}$$

which is what we obtain by substituting 'number-of-planets(x)' for '$\phi(x)$' and '$\Box(x > 7)$' for '$\Psi(x)$'.

The question now is, is A′ actually true? The answer is: obviously it is. The witness to '$(\exists x)$' is clearly 9, for 9 satisfies all the conjuncts. So A is true as well!

This is the answer modal sceptics want, since it appears to show that if one uses modal apparatus one is eventually compelled to accept as true many assertions that those who believe in the integrity of modal apparatus would insist are false.

The other possibility is that "Necessarily the number of planets is greater than 7" is of the form $\Box\Phi$ (where Φ is $\phi((\iota x)(\psi(x)))$). Ever since Kripke it has been almost universally felt that an expression purportedly of the form

$$\Box\Psi$$

is really of the form

$$(\forall\mathcal{W})(\mathcal{W} \models \Psi)$$

where the fragment '$(\forall\mathcal{W})$' is a quantifier over entities called 'possible worlds'.

According to this view,

A: $\Box((\iota x)(\text{number-of-planets}(x)) > 7)$

is really (unravel the box)

A″: $(\forall\mathcal{W})(\mathcal{W} \models ((\iota x)(\text{number-of-planets}(x)) > 7))$

Next we unravel the singular description by appealing to theorem 2.1. Using the notation of that theorem we find ϕ is number-of-planets(x) and Ψ is $(\forall\mathcal{W})(\mathcal{W} \models y > 7))$. Now we have to check that no variable free in "number-of-planets(x)" is bound by any quantifier in "$(\forall\mathcal{W})(\mathcal{W} \models y > 7)$". Now believers in possible worlds regard all atomic formulæ as having an extra place which is occupied by a variable over possible worlds: primitive assertions cannot be "x is F" but x is F at $\mathcal{W}$". Therefore "number-of-planets(x)" contains a hidden variable over possible worlds, (so that it is something like "$\mathcal{W} \models$ number-of-planets(x)") and so the variable-capture condition of the theorem is *not* satisfied. If this is so, then the correct way to interpret A in the language without singular terms is

$$(\forall\mathcal{W})(\mathcal{W} \models (\exists x) \bigwedge \begin{cases} \text{number-of-planets}(x) \\ (\forall y)(\text{number-of-planets}(y) \to y = x)\,) \\ x > 7 \end{cases}$$

which states that in every possible world there are more than seven planets. Modal realists are united in agreeing that this is false.

It seems fairly clear to us nowadays that—since it gives us an analysis of the troublesome proposition that comes out false as desired—this is in some sense the "correct" insight to apply to Quine's puzzle. That is to say: it is the correct insight to apply to the puzzle if we are going to take the puzzle seriously—that is to say, not repudiate the modal language altogether. Quine's repudiation of modal talk is very germane to the wider issues with which we are concerned here, for his analysis of modal language (in [1966b]) famously allows modal operators attached to closed formulæ as syntactic sugar for expressions which name those formulæ and assert that they are logically valid. Expressions consisting of modal operators attached to the front of other formulæ are not generated in this way and are not justified by this analysis. Whether or not one is attracted by this analysis is not the point here: I am bringing it up because it is another example of an analysis of the kind I wish to provide later for ordinal arithmetic.

Quine's puzzle at least provides us with an example to illustrate the fact that the rewriting rules that arise from interpretations need not be confluent.

2.3.1 *Carnap's treatment*

However, there is another approach to this, due to Carnap, that (as we would now put it) swallows Quine's bait. It takes "the number of planets is necessarily greater than seven" to be of the logical form of a predicate attached to a singular term, and yet avoids the unpleasant conclusion.

The idea is to have a much stronger notion of equality, so that fewer things are equal to each other, and it becomes correspondingly harder to find a witness to the existential quantifier. If we do this sufficiently thoroughly, then the expression will come out false as desired.

Carnap's way of doing this (in [C1]) is to fragment all the old entities (9, which is also the number of the planets, the sum of 5 and 4 etc) into lots of new ones, so that practically nothing is identical to anything else. In particular there is now no entity that is both the number of planets and is necessarily greater than 7. There is a number 9, and there is another intensionally identified entity which is the number of the planets. Although there is a new equivalence relation $\sim$ that holds between these new entities, it is strictly weaker than equality, so there is, as we wanted, no witness to A′. What *is* true is that

$$(\exists x) \bigwedge \begin{cases} \text{number-of-planets}(x) \\ (\forall y)(\text{number-of-planets}(y) \to y \sim x)\,) \\ \Box(x > 7) \end{cases}$$

but since this has '$\sim$' instead of '=' it is not a proper translation of A.

Although many (including me) feel that Carnap's approach to this puzzle is not helpful, the idea it contains that one can clarify matters by paying close attention to the notion of equality involved in a problematic singular description is a useful one to which we will return in later sections.

Quine says something like this in his appendix to Carnap [1947].

Chapter 3

Virtual Objects

I warned the reader earlier that I am going to consider *only* virtual objects that arise from congruence relations. Although—as indicated—this is certainly not the only kind of virtual entity one might wish to think about it is at least a sensible place to start. Readers may feel that this simplification is an oversimplification, and may be reminded of the joke about the physicist who was consulted by a punter for advice on how physics could help him work out which horse to back. The physicist's research led initially to a basic theory which worked "only for a spherical horse travelling in a vacuum". But one has to start somewhere! Restricting attention to interpretations arising from congruence relations makes a treatment of the logical issues involved relatively simple. For another, at least some of the real-life reductionist positions belonging to the mainstream rely on equivalence relations: although the formulation "chemistry is the study of that part of physics for which the relations between atoms of having-the-same-number-of-protons is a congruence relation" might not be mainstream, its content is close to it. Finally, ordinals arise from an equivalence relation on wellorders, and our main concern here is with that case. Accordingly restricting our attention to virtual entities that arise from equivalence relations is not unreasonable. The time has now come to develop what I earlier (p. 13) called the **canonical simulation**.

3.1 Congruence relations

Fix a language $\mathcal{L}$ until further notice. It has predicate letters, function letters and constants. We may introduce extra function symbols by means of the manœuvres in section 2.2. It has countably many variables each of which is a lower case Roman letter with a subscript from $\mathbb{N}$, which will be omitted if it is zero.

We will need to define *congruence relations*. In what follows $\sim$ will of course be an equivalence relation. In most interesting cases the congruence relation will be definable, i.e., captured by a formula of $\mathcal{L}$.

Definition 3.1 (According to a theory T) an equivalence relation $\sim$ is a **congruence relation for**

- a predicate P iff $T \vdash ((\bigwedge_{i<n} x_i \sim y_i) \wedge P(\vec{x})) \to P(\vec{y})$ where n is the number of free variables in P;
- an n-ary function f iff $T \vdash (\bigwedge_{i<n} x_i \sim y_i) \to f(\vec{x}) \sim f(\vec{y})$.

P and f are not assumed to be atomic, and in general they won't be. The congruence relation will give rise to equivalence classes. If we write our congruence relation with a tilde: '$\sim$' then the equivalence class of x is customarily denoted by '$[x]_\sim$'. In contexts where it is clear which congruence relation we have in mind the tilde will be omitted.

This definition has to be given separately for predicates and for functions. Granted, one can eliminate function letters in favour of predicate letters and singular descriptions, but the notion of congruence for functions is weaker that than for predicates. Consider congruence of integers mod 17. This is a congruence relation for the binary *functions* $\times$ and $+$, though not for the ternary relations '$x + y = z$' and '$x \times y = z$'. However, definition 3.1 is horn. This suggests an operation: input P and $\sim$; output $\bigcap\{S \supseteq P : \sim$ is a congruence relation for $S\}$. This is of course the relation $(\exists\vec{y})(\bigwedge_{i<n} x_i \sim y_i) \wedge P(\vec{y}))$. This construction is useful if—as with congruence mod 17—we are working in a context where we have an equivalence relation and there is a relation of importance to us for which the equivalence relation is *not* a congruence relation, and we would like it to be. We plump out the relation $\{\langle x, y, z\rangle : x + y = z\}$ to the relation $\{\langle x, y, z\rangle : 17|(x + y - z)\}$, and congruence mod 17 really is a congruence relation for this last predicate. This plumped-out relation is functional in the sense that for all x and y there is a z such that $17|(x + y - z)$ and although this z is not unique all such z are congruent mod 17, as follows

$$(\exists z)(x + y = z \ (\text{mod } 17) \wedge (\forall z')(x + y = z' \ (\text{mod } 17) \to z \sim_{17} z'))$$

which will in due course justify the use of a singular term (namely $x + y$ (mod 17)) along the lines of $(\iota z)(x + y = z \ (\text{mod } 17))$ in the language with variables ranging over integers mod 17. This is of course the way to represent functions as definite descriptions with free variables. Notice the occurrence of '$\sim_{17}$' instead of '=' in the formula above: in other respects the displayed formula matches the template of definition 2.1 exactly.

The set of equivalence relations that are congruence relations for P is a chain-complete poset. If $\sim$ is a congruence relation for P, so is any equivalence relation that is stricter than $\sim$. This is not the case for congruence relations for functions! The assertion that $\sim$ is a congruence relation for P, and that it is a congruence relation for f are both horn, but there is an important difference in that the single unnegated atomic formula in the body of the horn clause in the second case involves $\sim$ and in the first it doesn't. This has the effect that any function gives rise to a closure operation on equivalence relations: if f is a binary function and $\sim$ likens a to a' and b to b' then it should also liken $f(a, b)$ to $f(a', b')$.

3.2 Extending the language

So we have an equivalence relation, written '$\sim$', which is a congruence relation for various predicates P_i, P_2 ... where P_i is of arity n_i.

Let us now devise a new language, $\mathcal{L}^*$. $\mathcal{L}^*$ will have a new class of variables which will be lower case Greek letters: $\alpha, \alpha_1, \alpha_2, \ldots \beta, \beta_1, \beta_2 \ldots$ corresponding canonically to the Roman letters a, b of $\mathcal{L}$, so that there is a function $\mathcal{I}$ that sends Greek letters $\alpha, \alpha_1, \alpha_2, \ldots \beta, \beta_1, \beta_2 \ldots$ to Roman letters $a, a_1, a_2 \ldots$, $b, b_1, b_2, \ldots$, $c, c_1 \ldots$, ... with which we began. We do this by sending each Greek variable to the corresponding Roman variable with the same subscript. The new (Greek) variables in $\mathcal{L}^*$ will be used so that they appear to range over $\sim$-equivalence classes but in fact are ontologically cost-free.[1]

For each old such predicate letter P_i (with n_i arguments) we supply $\mathcal{L}^*$ with a new predicate letter P_i^*, also with n_i arguments. The difference between P_i and the corresponding P_i^* has ghosts in natural languages. Take for example the binary relation "fewer than". We say "Five smoked salmon is fewer than ten smoked salmon" but "five is less than ten" (not: "five is fewer than ten"). "The temperature is hotter today" sounds wrong: the *day* is hotter: the temperature is *higher*. Function letters are treated similarly. We will have an interpretation $\mathcal{I}$ that sends every Greek variable to the Roman variable with the same subscript, and send the predicate letter P_i^* to P_i.

[1]In *From a Logical Point of View* p 117 (section 4) Quine makes the point that it is always possible in circumstances like this to quantify (apparently) over equivalence classes by reinterpreting the congruence relation as '='. This may excite objections on the grounds that '=' is not an ordinary predicate to be reinterpreted *ad libitum* but is part of the logical vocabulary. It is out of respect for this objection that we use the special word 'simulations' for those interpretatations that treat '=' as a predicate letter instead of as a part of the logical vocabulary.

A picture might help.

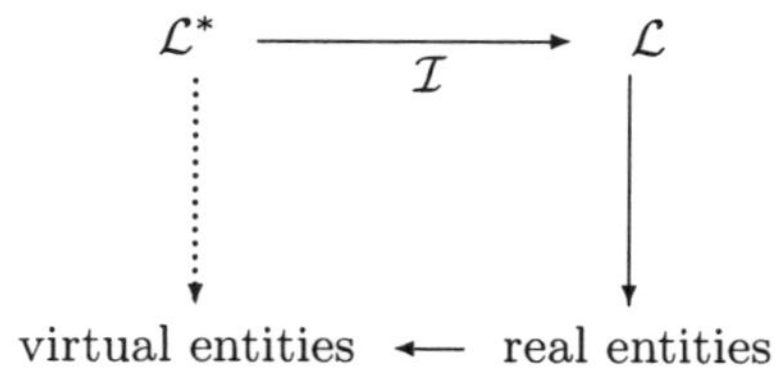

The top row contains syntax, and the bottom row contains things. The vertically downward arrows are interpretations as in model theory. The left-to-right arrow on the top line is an interpretation in the sense of this tract. The right-to-left arrow on the bottom line is the quotient map. The point is that composition of arrows make sense: the function $\mathcal{I}$ can be composed with the interpretations and assignment functions used in the semantics of $\mathcal{L}$ to give an exactly analogous semantics for $\mathcal{L}^*$... and the diagram commutes. Virtualism allows one to think that the semantics arises from the dotted arrows whereas really it arises from the solid arrows.

This is pure and simple, and easy to understand, and so a good place to start from. However it will usually be insufficient because we will want to talk about the new entities and the old in the same breath. What we really want is a map from $\mathcal{L} \cup \mathcal{L}^*$ into $\mathcal{L}$. It will turn out that the effort involved in doing this gets us also a map from a language slightly larger than $\mathcal{L} \cup \mathcal{L}^*$. We can even introduce at no cost a binary relation symbol and a function symbol. The binary relation symbol is $\mathcal{B}$ which has a Roman variable to its left and a Greek variable to its right: $a \mathcal{B} \beta$ means that a belongs to (the—as it were—equivalence class) β. The function symbol will be written "$\alpha = [b]_\sim$" and read "α is the $\sim$-equivalence class of b". Evidently $[a]_\sim$ is a singular term in the obvious sense: it's an abbreviation of '$(\iota\beta)(a \mathcal{B} \beta)$', and it is legitimate because $\mathcal{I}$ of $(\exists\beta)(a \mathcal{B} \beta) \wedge (\forall\beta')(a \mathcal{B} \beta' \to \beta = \beta'))$' is $(\exists b)(a \sim b \wedge (\forall b')(a \sim b' \to b \sim b')$ which follows from $\sim$ being an equivalence relation. Finally we have an equality, $=$. This is a *two-sorted* language, in that every variable must be of precisely one flavour of the two, Greek and Roman, and Greek variables cannot be placed as arguments to the old predicate letters P_i, nor Roman letters placed as arguments to the new predicate letters P_i^*. As my late colleague John Iorns once put it "No puns!".

Let us call this new language $\mathcal{L}^{**}$, or—since we will not be spending any time on $\mathcal{L}^*$ and so can recycle its name—$\mathcal{L}^*$ again.

We now need to show how closed formulæ (sentences) in $\mathcal{L}^*$ can be translated into the old language $\mathcal{L}$ containing only the old Roman variables and the original predicates $P_i : i \in I$. As before we will use the letter '$\mathcal{I}$' for this interpretation as well as for the map from Greek variables to Roman variables.

There is a slight niggle in that we can no longer send a Greek letter to the Roman letter with the same subscript. The Roman variable is already there and we need the map on variables to be injective. We will dine at Hilbert's Hotel. We will send each Greek variable to the corresponding Roman variable with the subscript *doubled*, and we send each Roman variable with subscript n to the Roman variable with the same body and with subscript $2n+1$.

Definition 3.2 The canonical simulation
Define $\mathcal{I} : \mathcal{L}^* \to \mathcal{L}$ by recursion as follows.

(1) $\mathcal{I}$ of '$=$' between Greek variables is '$\sim$';
(2) $\mathcal{I}$ of '$=$' between Roman variables is '$=$';
(3) $\mathcal{I}$ of $\ulcorner P_i^* \urcorner$ is $\ulcorner P_i \urcorner$;
(4) $\mathcal{I}$ sends Roman predicate letters to themselves;
(5) $\mathcal{I}$ of a Greek variable is the corresponding Roman variable with the subscript doubled;
(6) $\mathcal{I}$ of a Roman variable with subscript n is the Roman variable with the same body and subscript $2n+1$;
(7) $\mathcal{I}$ of $\ulcorner a_i \, \mathcal{B} \, \beta_j \urcorner$ is $\ulcorner a_{2i+1} \sim b_{2j} \urcorner$;
(8) $\mathcal{I}$ of $\ulcorner \alpha_i = [b_j]_\sim \urcorner$ is to be $\ulcorner a_{2i} \sim b_{2j+1} \urcorner$;
(9) $\mathcal{I}$ sends quantifiers to themselves.

If T is a theory in $\mathcal{L}$ then clearly $\mathcal{I}^{-1}$"T is a theory in $\mathcal{L}^*$, and is a conservative extension of T. If we already have a natural deduction formulation for T we can extend it to a natural deduction formulation for the corresponding theory $\mathcal{I}^{-1}$"T in $\mathcal{L}$ by adding, for each new predicate P_i^*, a pair of a P_i^*-introduction and a P_i^*-elimination rule:

$$\frac{P_i(a_1 \ldots a_n)}{P_i^*(\alpha_1 \ldots \alpha_n)} \qquad \frac{P_i^*(\alpha_1 \ldots \alpha_n)}{P_i(a_1 \ldots a_n)}$$

If we do this we will find that $\mathcal{I}$, considered as a map defined on $\mathcal{L}^*$-proofs, takes values among $\mathcal{L}$-proofs. The above rules give rise to derived rules

$$\frac{\mathcal{I}(\Phi)}{\Phi} \qquad \frac{\Phi}{\mathcal{I}(\Phi)}$$

There are obvious generalisations of this to the cases where $\mathcal{L}^*$ is many-sorted, and there is more than one congruence relation in play, one for each sort. While we are about it, let's nail down a name for this phenomenon of having congruence relations for each sort.

Definition 3.3 Let ϕ be a predicate letter with free variables $\vec{x}$ where, for each $i \in I$, x_i is of type σ_i. For each $i \in I$ let $\sim_i$ be an equivalence relation on things of type σ_i. If $(\forall\vec{x})(\forall\vec{y})((\phi(\vec{x}) \wedge \bigwedge_{i\in I} x_i \sim_i y_i) \to \phi(\vec{y}))$ then we say that the bundle $\{\sim_i : i \in I\}$ is a **suite-of-congruence-relations** for ϕ.

In slang: wiggle each variable by the equivalence relation of the appropriate level.

One would expect this elaboration of $\mathcal{L}^*$ from $\mathcal{L}$ to be model-theoretically neat, and so it turns out. Any model of a $\mathcal{L}^*$ theory $\mathcal{I}^{-1}``T$ has a reduction that is a model of the corresponding $\mathcal{L}$-theory $T$. Any model $\mathfrak{M}$ of an $\mathcal{L}$-theory $T$ gives rise to a model $\mathfrak{M}'$ of the $\mathcal{L}^*$-theory $\mathcal{I}^{-1}``T$. Let M be the domain of $\mathfrak{M}$. Then domain of $\mathfrak{M}'$ is the union of two disjoint copies M_1 and M_2 of M: everything in M appears twice: once as itself, and once as a duplicate destined to be the designation of Greek variables. (This is foreshadowed in the way we had to double subscripts of Roman variables when defining $\mathcal{I}$.) Roman variables range over things in M_1; Greek variables range over things in M_2. The graph of the relation $\mathcal{B}$ is the set $\{\langle x, y\rangle : x \in M_1 \wedge y \in M_2 \wedge x \sim y\}$. M_1 and M_2 have different concepts of equality. In M_1 the concept of equality is the same as in $\mathfrak{M}$; in M_2 two elements are equal iff they were related by $\sim$ in $\mathfrak{M}$. It is not too fanciful—and may even be helpful—to think of there being only *one* copy of M, but two ways of thinking of equality between its elements. Two things could be unequal *qua* members of $\mathfrak{M}$ but equal as equivalence classes (as it were). We will return to this theme later.

One thing is clear already. If, for some range of entities, we have a theory about them which we feel is adequate to the expressive demands of those entities, and this theory is of the form $\mathcal{I}^{-1}``T = T^*$ for some $T \subseteq \mathcal{L}$ then we have semantics for T^* which alludes only to entities of the kind denoted by expressions of $\mathcal{L}$: $\mathfrak{M}'$ contains only things in $\mathfrak{M}$, in duplicate or more perhaps, but it doesn't contain anything new. We could speak of this as *virtualism* for the entities described by the theory T^*.

I would in principle agree with the reader who says there are too many 'ism' words already, and isn't 'reductionism' enough to be going on with? Well, this is a special case of reductionism, where the reduction arises from a congruence relation or family of congruence relations. It's the simplest place to start and it deserves a name.

3.2.1 *Implementations of languages with a canonical simulation*

In this chapter we have been considering languages that arise from a language $\mathcal{L}$ by considering congruence relations in $\mathcal{L}$. In these circumstances there is a canonical simulation, as we have seen. Where there is a canonical simulation, it is natural to think about implementations that respect it in the following sense. Recall that an interpretation $\mathcal{I}$ is an implementation iff $\mathcal{I}$ of equality is equality. A little thought will reveal that it can send equality to equality only if it it incorporates a total function f satisfying $(\forall xy)(x \sim y \longleftrightarrow f(x) = f(y))$, and interprets quantifiers $(\exists \alpha_i)(\ldots)$ as $(\exists a_{2i} \in f``V)(\ldots)$. (*i.e.*, quantifiers over Greek variables arise only from quantifiers over Roman variables that are restricted to the range of f.)

Typically we might also want a sort of *representation* condition, namely $(\forall x)(x \sim f(x))$. This might appear to be an obvious absolute requirement, but this turns out not to be so. It seems natural to speak of $f(x)$ as the *representative* of x, but if the representation condition is not met, perhaps it would be better to say $f(x)$ is the *prefect* of x ('prefect' as in 'préfecture'—prefects are *not* elected by the inhabitants of the préfecture from among their number, but are appointed by the central government).

We say that an implementation $\mathcal{I}$ is **faithful** iff it satisfies this representation condition. That is to say, if it picks a representative from each equivalence class.

A weaker condition that is often adequate to our purposes is one that says that for each predicate P for which $\sim$ is a congruence condition we have another predicate P' such that $(\forall x_1 \ldots x_n)(P(x_1 \ldots x_n) \longleftrightarrow P'(f(x_1) \ldots f(x_n)))$. (In realistic cases like the one to follow the map that sends P to P' is primitive recursive in the Gödel numbers.)

An illustration of the dispensibility of the representation condition is the Von Neumann implementation of ordinal universally used in wellfounded set theory. A Von Neumann ordinal is not a wellordering, even though it has a wellordering—indeed a canonical wellordering. Although the Von Neumann implementation of ordinals in ZF does not satisfy the representation condition and therefore is not faithful, it does satisfy the weaker condition of the last paragraph, and it would satisfy the stronger one if we tweaked it slightly to take an ordinal to be not a transitive set wellordered by $\in$ but rather the restriction of $\in$ to such a set.[2] This is in contrast to the Von Neumann implementation of cardinals in ZFC, where the cardinal α really is a set of size α, so the implementation really is faithful. In fact the Von Neumann implementation of cardinals in ZFC does more still. For cardinals $\alpha \leq \beta$ we not only have an injection from $f(\alpha)$ into $f(\beta)$, the injection is actually the inclusion map: $(\forall \alpha\beta)(\alpha \leq \beta \longleftrightarrow f(\alpha) \subseteq f(\beta))$. (Remember that $\subseteq$ is *not* the relation between sets from which $\leq$ on cardinals arises: $\leq$ arises from

[2]Actually if we take a wellordering to be not a set of ordered pairs but the set of domains of its initial segments then the wellordering of a Von Neumann ordinal turns out to be the von Neumann ordinal itself, so on that view of wellordering the Von Neumann ordinals are faithful after all! We will exploit this fact on p. 67.

the plumped out (see page 28) version: $\{\langle x, y\rangle : (\exists x')(x \sim x' \subseteq y)\}$ which is the same as $\{\langle x, y\rangle : (\exists y')(y \sim y' \supseteq x)\}$. (Another fact that distracted me for a while!)). The existence of an implementation satisfying this extra condition is a nontrivial consequence of the axiom of choice. If every set of cardinals has an order-preserving set of representatives in this way then every Dedekind-infinite set has a countably infinite subset. However it is a theorem of Truss [1973] that every *finite* set of cardinals admits an order-preserving set of representatives in this way.

We have said nothing about whether the graph of f is to be a set, locally a set or even perhaps *setlike* (a property we will see in definition 5.5 which has the effect that if x is a set of sets, then the collection of cardinals of members of x is also a set) or none of the above. This is being mentioned at this stage merely to reassure the reader that—although it is being left open—it is not being forgotten.

3.2.1.1 *Subversion*

Implementations bring with them the possibility of puns. This is because the type distinctions that separate Greek from Roman variables can be subverted by the implementation. Any occurrence of a Greek letter 'α_i' in a place where a Roman letter is required can be replaced by '$f(\alpha_i)$'. Thus it becomes possible to ask questions that sound like "$3 \in 5$?", namely $f(3) \in f(5)$?

Subversion of this kind is well-known in Quine's NF. Normally the axiom of separation says only that if A is a set so is $A \cap \{x : \phi\}$ as long as ϕ is stratified. This restriction seems to keep the system consistent. However, when A is small in a suitable sense ("strongly cantorian") this syntactic restriction can be subverted in a manner strikingly like that of the previous paragraph.

3.2.2 *An illustration: utilities*

Let us have a simple example: utilities as virtual objects. Utilities arise from preference relations, which are (among other things) transitive and reflexive. Binary relations that are transitive and reflexive are **preorders**. Each preorder gives rise to a partial order and an equivalence relation in a canonical way. Let $\leq$ be a preorder. The two relations

$$\sim = \{\langle x, y\rangle : x \leq y \leq x\}$$

and

$$\leq^* = \{\langle [x]_\sim, [y]_\sim\rangle : x \leq y\}$$

are an equivalence relation and a partial order respectively. The domain of the partial order is the set of $\sim$-equivalence classes of the original domain. As is customary, I have written '$[x]_\sim$' for the equivalence class of x under $\sim$, and will—as is also customary—omit the subscript where understanding is not threatened.

Let us now define languages $\mathcal{L}$ and $\mathcal{L}^*$. $\mathcal{L}$ contains lower case Roman letters for variables, '=' for equality, a binary predicate letter $\leq$ and finally '$\sim$' a binary predicate letter that will be interpreted by an equivalence relation. $\mathcal{L}^*$ contains in addition lower case Greek letters and a binary predicate letter $\leq^*$. We will define an interpretation $\mathcal{I}$ of $\mathcal{L}^*$ into $\mathcal{L}$ in the style of definition 3.2. Thus

Definition 3.4

(1) $\mathcal{I}$ of a Greek variable is the corresponding Roman variable with the subscript doubled;
(2) $\mathcal{I}$ of '=' is '$\sim$';
(3) (iii) $\mathcal{I}$ of '$\leq^*$' is '$\leq$'.

So, for example, $\mathcal{I}$ of

$$(\forall\alpha)(\exists\beta)(\alpha \leq^* \beta \wedge \neg(\beta \leq^* \alpha))$$

is

$$(\forall a)(\exists b)(a \leq b \wedge \neg(b \leq a))$$

A typical and natural example of a preorder is an individual preference relation. The domain of the relation consists of *goods* (to use a piece of economists' jargon). *Prima facie* the only assumption we should make about a preference relation is that it is transitive and reflexive, that is to say, that it is a preorder. One shouldn't expect it to be trichotomous because one may be genuinely incapable of choosing between a bottle of burgundy and a bottle of claret, but the assumption that preference relations are trichotomous is a simplifying assumption which is legitimate when the topic is being developed purely for the sake of illustration.

What we have here is a language that purports to speak of range of (virtual) objects for which the expression "values" leaps to mind. Nevertheless we are quantifying only over objects and not over the utilities we ascribe to them. This is the simplest and most natural example known to me of a range of virtual objects arising from an equivalence relation.

3.2.2.1 *Baskets of goods*

Of course the domains of realistic preference relations typically have a lot more structure than this. A good (type)[3] need not be merely a thing like—say—the type of apples (one token of which is an apple) but could also be a thing like—to pull a notation out of the air—apple $\oplus$ apple, a token of which is two apples. A *basket* of goods is also a good. We would naturally expect $\oplus$ to be monotone. That is to say we expect $(\forall abcd)((a \leq b \wedge c \leq d) \rightarrow (a \oplus c \leq b \oplus d))$. There is also the possibility of a null good, notated '0', satisfying the obvious condition $(\forall x)(x \leq (x \oplus 0) \leq x)$. For a more detailed discussion of operations on objects see Link [L2].

Naturally we will want to expand $\mathcal{L}^*$ to a language containing a symbol—'+', say—such that $\mathcal{I}(+) = \oplus$.

Once we have done this we have a theory of things that one might call *utilities* which have an additive structure and a total ordering. Perfect sense can be made of expressions in this language which contain quantifiers apparently over things which can only be utilities. We can do this even if we do not suppose the world to contain anything other than mere goods.

We can spin this story out by considering the possibility of goods of negative value, and eventually obtain a theory of value as a module over Z (or even an account of $\Re$) but that would not further serve the purpose of illustration which is the sole function of this example here. We will content ourselves with the slogan-style virtualist summary:

The theory of value is the study of those relations between goods for which being-equally-desirable is a congruence relation.

See Suppes [1957], [1959] and Krantz [1971] on this.

3.2.3 *Some standard mathematical examples*

Although it is Philosophy is full of claims that *A*s are just logical constructions out of *B*s, it is to Mathematics that we must look for tangible illustrations of such logical constructions.

Let us consider the construction of the (positive and negative) integers from the naturals. Given two naturals a and b we seek x which will be a solution to the equation $a + x = b$. If $b \geq a$ this will have a solution in $\mathbb{N}$, but not otherwise. If x is a solution to $a + x = b$ then, for all c, it will be solution to $(a + c) + x = (b + c)$, so the problem posed by the pair $\langle a, b \rangle$ is in some sense the same as the problem posed by any pair $\langle a + c, b + c \rangle$. It's easy to check that two pairs $\langle a, b \rangle$ and $\langle m, n \rangle$ pose the same problem (and have the same solution) iff $a + n = b + m$, so we can take the solution to $a + x = b$ to be the set $\{\langle m, n \rangle : a + n = b + m\}$. If we do this even

[3] Here I mean 'type' as in the type-token distinction.

for equations $a + x = b$ that already have solutions then we have a set of things on which we can define addition and subtraction and contains all the natural numbers, or at least duplicates of them.

There are various striking differences between this development and the kind of development that would be involved in a logical-constructions-out-of-sense-data development.

The first is that analyses of this kind seem to have no ontological *bite*: there is no suggestion that the construction above shows that integers are not real objects even though the naturals are, or that it is possible to do all the mathematics that needs integers while believing only in $\mathbb{N}$. Anybody who has got that far is already up to their navel in abstract entities and has no virtue left to protect. There is nothing *at stake*. The purpose of constructions like these is to present the new mathematical entities in a way that explains why we need to believe in them.

Another difference is that with these analyses that are conducted entirely inside the mathematical context, there is usually no question mark over the existence of the equivalence classes (and therefore the quotients) modulo the congruence relations. (There was a time when there were people who thought that perhaps the collection of all reals—or even perhaps the collection of all naturals—was not a set (which would have had the effect that sethood of the equivalence classes would have been very hard to prove) but these concerns have now evaporated, and were in any case, not really concerns about set existence at all but about something much less cut-and-dried.)

Finally in cases like the construction of the integers from the naturals that we have just seen, nobody in their right mind would wish to have any implementation of integer arithmetic back into the theory of ordered pairs of naturals that was faithful. If one wished to be concrete about it, one would identify the integer $-n$ with the set of ordered pairs $\{\langle x, y\rangle : x = y+n\}$, but certainly not with any one ordered pair in it. To do that would be somehow to regard one of the equivalent equations as being more important than the others, and none of them are.

A development like this one allows what model theorists call *elimination of imaginaries*. See Hodges [1993] pp 157ff.

3.3 Second-order and higher-order theories

The first-order virtual language arose as a way of conveniently notating facts about relations between things for which the equivalence relation $\sim$ was a congruence relation. The time has now come to think about a language in which we can express facts about sets of things for which the equivalence relation in hand is a congruence relation in the appropriate sense. Just as we did in the first-order case, we will develop this machinery purely for the virtual entities in the first instance.

The higher-order language is of course two-sorted, so we will need the apparatus of *suites of congruence relations* of definition 3.3

Let us suppose that $\mathcal{L}$ had upper-case Roman letters for ranging over sets of whatever it was that the lower-case Roman letters ranged over. $\mathcal{L}^*$ will of course have upper-case Greek letters to correspond to them. $\mathcal{L}$ also had $\in$, which had a lower-case Roman letter to its left and an upper-case Roman letter to its right. $\mathcal{L}^*$ will have a corresponding membership relation, which we will write '$\in_1$' and an equality predicate $=_1$ which sits between two upper-case Greek letters. $\in_1$ is of course to be extensional (it's set membership after all) and we know how to interpret in $\mathcal{L}$ the equality between lower-case Greek letters. We must have

$$(\forall \Theta, \Delta)(\Theta =_1 \Delta \longleftrightarrow (\forall \alpha)(\alpha \in_1 \Theta \longleftrightarrow \alpha \in_1 \Delta))$$

This actually tells us what the congruence relation will be from which the equality symbol between upper-case Greek letters will arise.

First we define an equivalence relation in $\mathcal{L}$ as follows.

Definition 3.5 $S \sim^+ T \longleftrightarrow (\forall x \in S)(\exists y \in T)(x \sim y) \wedge (\forall y \in T)(\exists x \in S)(x \sim y)$

This is actually a standard way of using an equivalence relation on a set X to give an equivalence relation on $\mathcal{P}(X)$, and this notation has been used for this before. I believe it is due to Roland Hinnion.

Let us write '$\sim^+$' as '$\sim_1$', with a view to having larger subscripts from $\mathbb{N}$ later for predicates on objects of higher types.

We can now see that $=_1$ has to be sent to $\sim_1$, Also '$\alpha \in_1 \Gamma$' will have to be sent to something like '$a \in G$'. In fact it will have to be sent to the plumped-out version of this relation which is '$(\exists a' \sim a)(a' \in G)$' or equivalently '$(\exists G')(a \in G' \sim_1 G)$'. So $\{\sim, \mathcal{I}(=_1)\}$ must be a suite-of-congruence-relations for $\in_1$.

Now, just as in the first-order case, whenever ϕ is a second-order predicate of $\mathcal{L}$ for which $\{\sim, \sim_1\}$ is a suite-of-congruence-relations, we can introduce a new predicate ϕ^* of the same arity as ϕ and we naturally send '$\phi^*(\vec{\Theta})$' to '$\phi(\vec{T})$'.

As before, we are more likely to be interested in the combination language with both Greek and Roman variables than just the new language. To interpret this new combined second-order language in the old second-order language $\mathcal{L}$ we add to definition 3.2 the clauses:

Definition 3.6

10 Upper-case Greek letters get sent to the corresponding uppercase Roman letter with the subscript doubled;
11 Upper-case Roman letters with subscript i get sent to the variable with the same body and the subscript $2i + 1$;
12 $\mathcal{I}$ of $\ulcorner \Gamma_i =_1 \Theta_i \urcorner$ is $\ulcorner G_{2i} \sim_1 T_{2j} \urcorner$;
13 $\mathcal{I}$ of $\ulcorner \alpha_i \in_1 \Gamma_j \urcorner$ is $\ulcorner (\exists a_{2i+1} \sim a_{2i})(a_{2i+1} \in G_{2j}) \urcorner$;
14 $\mathcal{I}$ of $\ulcorner \phi^*(\vec{\Theta}) \urcorner$ is $\ulcorner \phi(\vec{T}) \urcorner$.

We can add clauses for higher-order analogues of $\mathcal{B}$ and the square bracket notation too if we wish.

This definition makes distinctions which we will later collapse. It does not assume that sets of whatever it is that the Roman variables of the original languge $\mathcal{L}$ ranges over can be sets of those same things. In the first special case of interest to us—cardinal numbers as virtual entities over sets—we find that sets of sets are of course sets again, so that the distinction between lower and upper case Roman variables is not needed. The distinction between lower and upper-case *Greek* variables will remain, however: a set of cardinals is not a cardinal, except in a pun!

So far we have only considered how to interpret talk about *sets* of virtual entities, not *multisets* or *lists*. No doubt something general can be said about relations in the base language from which talk about multisets can arise, but in the only case that we really are compelled to treat (multisets of cardinals see page 45) talk of multisets arises in a rather pleasing and neat way which would certainly be used to override anything more general one could do here.

3.3.1 *Third- and higher-order virtual entities*

This process can be extended any finite number of times we like, so we may as well sketch the result of doing it infinitely often. It's quite hard to do this explicitly, as there is no orthographically natural second step of the process that took us in one step from lower-case Roman letters (for first-order variables) to capital Roman letters (for second-order variables). What are we to use for third-order variables? Shortage of alphabets and fonts is the single biggest obstacle to understanding and presenting this material.

Suppose $\mathcal{L}$ contains variables of all finite levels too, and a single polymorphic membership predicate letter that sits between nth order and $n+1$th order variables for each n.

Let the higher-order language $\mathcal{L}^*$ contain, for each $n \in \mathbb{N}$, a range of variables of level n. $\mathcal{L}^*$ will also contain predicate letters $=_n$ and $\in_n$ analogous to $=_1$ and $\in_1$, with the obvious axioms of extensionality and logical equality axioms.

$\mathcal{I}$ of $=_n$ will be $\sim_n$, which is defined by recursion as follows. $\sim_1$ has just been defined, and we set $\sim_{n+1}$ to be $(\sim_n)^+$.

$\mathcal{I}$ of $\in_{n+1}$ will be the relation $(\exists x' \sim_n x)(x' \in Y)$

And now, as before, we can add to $\mathcal{L}^*$ new predicate letters corresponding to old predicate letters in $\mathcal{L}$ for which the equivalence relations $\{\sim_n : n \in \mathbb{N}\}$ are a suite of congruence relations.

Adding more levels (as we are doing here) doesn't commit us to making more copies (as in the discussion on page 32). We still have only two. Later we will be considering constructions that require us to make multiple copies.

Chapter 4

Cardinal Arithmetic

Remarkably one will not find a definition of cardinal arithmetic in the places where one might expect to find it, such as books or survey articles on the subject (Holtz-Steffens-Weitz [1999] for example, or Shelah [1992]) so there is still scope for a definition to be hawked. Here is a sloganising suggestion:

> **Cardinal arithmetic is the study of those relations between sets for which equinumerosity is a congruence relation.**

... but of course it's not quite that simple, because we are going to treat higher-order cardinal arithmetic in this chapter, and the subject matter of higher-order cardinal arithmetic is that part of set theory for which the equivalence relations of equipollence, equipollence$^+$ etc... form a suite of congruence relations.

In the terminology introduced earlier, this is virtualism about cardinals. I am setting out not to defend or attack this doctrine but to explain it: it needs to be explained before either of those things than happen. Virtualism about cardinals is a *prima facie* attractive theory because it so strikingly respects our intuitions that puns like '$3 \in 5$' should be accorded groans rather than truth-values. Any account of cardinals that professes to have a good story about $3 \in 5$ and then tells a similar story about commutativity of addition probably has a suspect account of commutativity of addition as well. Any reader who feels that the significance of this point is slipping from their grasp need only perform the following thought-experiment. Hardy said of Ramanujan that "the natural numbers were his personal friends". Now imagine what Ramanujan would reply if one were to say to him "I've just made an exciting discovery that will be very useful to you: a natural number is a wellfounded hereditarily transitive finite set wellordered by $\in$".

In the virtualist account cardinal arithmetic is going to arise from set theory, so let's start with that.

4.1 The languages of set theory and arithmetic

We will need an underlying set theory in which to perform this interpretation. Since it is the interpretation we are interested in rather than the endeavour to prove as many theorems about cardinals as we can (which is the usual aim) I would like to say as little about the theory itself as possible. Any of the weak systems of Mathias [2001] will do.

Unfortunately (from the point of view of the expositor) one cannot develop a theory of equipollence without bijections and one cannot easily do that without having ordered pairs. This complication is what makes the cardinal arithmetic case different from the cases of modular arithmetic or the rescension of the integers from $\mathbb{N}$ which we considered earlier: in those cases no extra apparatus was needed.

There are several ways out of this quandary, and this variety of available approaches is a peculiarly unwelcome distraction, because it gives the impression that what we are trying to sort out is another implementation issue ("how do we implement bijections?") whereas all that we have to decide is which one of the various equivalent set theoretic formulæ that say that two things are the same size is the most convenient one to start from.

It is possible (a) to talk about bijections between sets without talking about ordered pairs at all, by means of branching quantifiers, thus

$$\begin{matrix}(\forall x \in A)(\exists y \in B) \\ (\forall z \in B)(\exists w \in A)\end{matrix} (x = w \longleftrightarrow y = z)$$

but I suspect few readers will find this to their taste. This is unfortunate, for although on the face of it this is certainly a bizarre way to procede, it is by far the neatest solution technically since it involves no extra gadgetry. A brief look at the other possible strategies will soften the reader's heart to option (a) as it softened mine.

Another solution (b) is to settle on an implementation of ordered pair once and for all, and then promptly forget it. This is certainly the most usual approach. It has the defect of not being "type-level", or **homogeneous**. However a more serious objection is that it is against the whole spirit of the programme here, which is to provide an analysis that is implementation-independent.

(There is also (b)$'$, an elegant construction of Henrard—never published—which expresses the presence of a bijection between sets A and B in terms of overlapping doubleton subsets of $A \cup B$. This has much the same effect as (b), but it does enable us to talk about bijections without invoking ordered pairs, even though it cannot handle arbitrary binary relations.)

A solution (c) which is in that spirit is to take pairing and unpairing as primitive constructors, and decide to respect a typing system according to which expressions like '$x = \mathtt{fst}(x)$' are ill-typed. This is in a sense what one *ought* to do, and some parts at least of this approach will have to be kept seaworthy because although we can dispense with ordered pairs in cardinal arithmetic (and even in ordinal arithmetic as well, as we shall see) we cannot expect to be able to pursue this kind of manipulation of relations in a more general setting without confronting ordered pairs sooner or later. But (c) is objectionable for two reasons: (i) it is complex and (ii) unless it is done very carefully there are nontrivial set-theoretic consequences, since if $V = V \times V$ the universe clearly cannot be finite and we don't want its infinitude to be provable by means of inelegant nonconservative reasoning that exploits linguistic add-ons. However the reason why I shall not go down that road is simply that it uses up too much potentially useful notation: earlier draughts of this essay have convinced me that any attempt to execute this approach would result in a document as unreadable as Russell and Whitehead's *Principia*. The extra syntactic cost is very like the cost of having ramifications as well as types. I shall *sketch* how to do things in an implementation-independent way, as one should—and may return to it from time to time, but I shall in fact mostly stick to (a). It is true that cartesian product ("$x \times y$", whence multiplication) and the set of all functions from x to y ("$x \to y$" whence exponentiation) need ordered pairs but when this fact raises its head I shall pretend I am doing (b). *Video meliora proboque; deteriora sequor.*

4.2 The canonical simulation

The formula that states that x and y are the same size:

$$\begin{matrix}(\forall x \in A)(\exists y \in B) \\ (\forall z \in B)(\exists w \in A)\end{matrix} (x = w \longleftrightarrow y = z) \tag{4.1}$$

is a bit of a mouthful and we will abbreviate it to $x \sim_c y$. Similarly $x \hookrightarrow_c y$ will abbreviate $(\exists y' \subseteq y)(x \sim_c y')$. Or do we mean maybe $(\exists x' \supseteq x)(x' \sim_c y)$? Actually it doesn't matter, as they can be shown equivalent in any of the weak set theories we would consider using: both are the plumped-out version of the relation $x \subseteq y$ with respect to the equivalence relation $\sim_c$.

This binary relation will give rise to the relation $\leq_c$ on cardinals. The subscript will be omitted where the omission will not cause ambiguity. In a similar sense $(\exists a' \sim_c a)(\exists b' \sim_c b)(a' \cap b' = \emptyset \wedge a' \cup b' = c)$ will give rise to cardinal addition, $a \sim_c b \times c$ will give rise to to cardinal multiplication, and $a \sim (b \to c)$ will give rise to cardinal exponentiation,

There is another binary relation for which $\sim_c$ is a congruence relation, namely "either x is empty or there is a surjection from y onto x". This gives rise to a relation between cardinals traditionally written '$\leq_*$'.

There are various other predicates considered in (large) cardinal theory which do, indeed, belong to cardinal arithmetic under this analysis. Clearly if x is a set with a countably complete nonprincipal ultrafilter on it and $x \sim_c x'$ then x' is also such a set. (The reader who does not know what a countably complete ultrafilter is need not panic—no further use will be made of this fact. The reader who does may wish to take note of the fact that $\{\sim_c, \sim_c^{++}\}$ is not a suite of congruence relations for "$\mathcal{U}$ is a countably complete non-principal ultrafilter on K", and that this is a situation that calls for equivalences defined in terms of permutations, as on page 46.)

We will deem that $\mathcal{L}$ includes all these new bits of syntax even though it started off just as a language with $\in$, pairing and unpairing.

We can now give the definition of the canonical simulation $\mathcal{I}$ of $\mathcal{L}^*$ in $\mathcal{L}$. Or rather, we consider the points of detail beyond definition 3.2 that it might be worth spelling out.

Definition 4.1

(1) $\mathcal{I}$ of '$\leq$' is '$\hookrightarrow_c$';
(2) $\mathcal{I}$ of $\ulcorner\alpha_i \leq_* \beta_j\urcorner$ is $\ulcorner a_{2i} = \emptyset \vee (\exists f)(f : b_{2j} \twoheadrightarrow a_{2i})\urcorner$;
(3) $\mathcal{I}$ of $\ulcorner\alpha_i + \beta_j = \gamma_k\urcorner$ is $\ulcorner(\exists a' \sim_c a_{2i})(\exists b' \sim_c b_{2j})(a' \cap b' = \emptyset \wedge a' \cup b' = c_{2k})\urcorner$;
(4) $\mathcal{I}$ of $\ulcorner\alpha_i = \beta_j \cdot \gamma_k\urcorner$ is $\ulcorner a_{2i} \sim_c b_{2j} \times c_{2k}\urcorner$;
(5) $\mathcal{I}$ of $\ulcorner\alpha_i = {\beta_j}^{\gamma_k}\urcorner$ is $\ulcorner a_{2i} \sim_c (c_{2k} \to b_{2j})\urcorner$;
(6) $\mathcal{I}$ of $\ulcorner|a_i| = \beta_j\urcorner$ is $\ulcorner a_{2i+1} \sim_c b_{2j}\urcorner$;
(7) $\mathcal{I}$ of all other uses of '$=$' is '$\sim_c$'.

To illustrate I shall remove the definite description from '$\alpha + \beta \leq \gamma$' and evaluate $\mathcal{I}$ of it. I shall do it the hard way, reading the equation as a singular description. (On the positive side, I shall omit subscripts with a view to improving readability)

'$\alpha + \beta$' is a singular description, so '$\alpha + \beta \leq \gamma$' is short for

$$(\iota\delta)(\alpha + \beta = \delta) \leq \gamma$$

which in turn is short for

$$(\exists\zeta)(\alpha + \beta = \zeta \wedge (\forall\zeta')(\alpha + \beta = \zeta' \to \zeta = \zeta') \wedge \zeta \leq \gamma)$$

$\mathcal{I}$ of '$\alpha+\beta = \zeta$' is $(\exists a'b'z')(a \sim_c a' \wedge b \sim_c b' \wedge z \sim_c z' \wedge z = a' \cup b' \wedge a' \cap b' = \emptyset)$, as in definition 4.1 clause 2. This then gives

$(\exists z)((\exists a'b'z')(a \sim_c a' \wedge b \sim_c b' \wedge z \sim_c z' \wedge z = a' \cup b' \wedge a' \cap b' = \emptyset) \wedge (\forall z'')((\exists a'b'z')(a \sim_c a' \wedge b \sim_c b' \wedge z'' \sim_c z' \wedge z'' = a \cup b \wedge a \cap b =$

$\emptyset) \rightarrow z \sim_c z') \wedge z \hookrightarrow_c c))$

Now since $\mathcal{I}$ of $z \hookrightarrow_c c$ is $(\exists z'''\exists c')(z''' \sim_c z \wedge c' \sim_c c \wedge z''' \subseteq c)$ in primitive notation this expands to

$$(\exists z)((\exists a'b'z')(a \sim_c a' \wedge b \sim_c b' \wedge z \sim_c z' \wedge z = a' \cup b' \wedge a' \cap b' = \emptyset) \wedge (\forall z'')((\exists a'b'z')(a \sim_c a' \wedge b \sim_c b' \wedge z'' \sim_c z' \wedge z'' = a \cup b \wedge a \cap b = \emptyset) \rightarrow z \sim_c z') \wedge (\exists z'''\exists c')(z''' \sim_c z \wedge c' \sim_c c \wedge z''' \subseteq c))$$

Eventually this will simplify to

$$(\exists a' \sim a)(\exists b' \sim b)(a' \cap b' = \emptyset \wedge a \cup b \subseteq c)$$

4.2.1 *Sets of cardinals*

We can invoke the apparatus of definition 3.6 to give us an interpretation back into set theory of talk of *sets* of cardinals as well. There are complications here over and above the treatment there, since in set theory there is no type distinction between sets and set-of-sets: upper-case Greek letters will be sent to lower case Roman letters. This means that there will now be *three* styles of variables to be shoehorned into the lower-case Roman letters instead of two, since upper-case Greek letters too will now be interpreteted by lower-case Roman letters. Subscripts will now have to be trebled and shifted not doubled and shifted. One way of avoiding this could be to contrive to think of a first-order set theory as actually a second order set theory, with the classes being exactly the same as the sets, so that $\mathcal{L}$ had upper-case Roman variables.

We can now handle suprema and infima of sets of cardinals as follows. $\{\sim_c, \sim^+\}$ is a suite-of congruence relations for "Every member of A injects into x and x injects into any other set with this property" and this will clearly give rise to a predicate in the language of cardinal arithmetic of being-the-sup of. Infima arise similarly. However, arbitrary sums and products of cardinals are operations on *multisets* of cardinals not sets of cardinals. Although this point is not often made we need to take it seriously here.

4.2.2 *Multisets of cardinals and the multiplicative axiom*

Of course one can think of multisets-of-cardinals as arising from sets-of-cardinals in the same way as multisets-of Xs usually arise from sets-of-Xs (usually taken to be functions from X to cardinals) but since we neet multisets only for sums of cardinals (in which circumstances we want all the members of the multiset to be disjoint) the clever thing to do is regard multisets of *cardinals* as arising from *partitions* (sets whose members are pairwise disjoint), and to take the equivalence relation between them (that is to be a component of the suite-of-congruence relations from which multisets will arise) to be $P_1 \sim^{+'} P_2$ iff there is a bijection between P_1 and P_2 which pairs off things of the same size. Then the sum of a multiset of cardinals arises from the binary relation "A is a partition and $\exists$ partition $A' \sim^{+'} A \wedge B = \bigcup A'$".

To deal with multisets we will need to add to $\mathcal{L}^*$ a set of variables in a new style, purporting to range over multisets of cardinals, and new function letters Σ and Π (for sum and product) which can be applied to these new variables, to be followed by '=' and then a lower-case Greek variable. There will be a membership relation symbol between lower-case Greek letters and letters in this new style. Since membership in a multiset has a multiplicity associated with it, rather than a mere simple truth-value there could even be a three-place predicate whose intended interpretation is "this cardinal belongs to this multiset with that multiplicity". In fact we will be compelled to do that if we wish to be able to encompass assertions like "The union of α disjoint sets each of size β is the same size as the union of β disjoint sets each of size α". This is a minefield, as we shall see.

Next we have to declare what the interpretation $\mathcal{I}$ will do to these new bits of syntax.

(1) $\mathcal{I}$ of a new ("multiset") variable will be a (lower-case) Roman variable restricted to range over partitions; (We now have *four* styles of variables to be shoehorned into lower-case Roman variables!)
(2) '$\Sigma(\Theta) = \alpha_i$' will go to '$\bigcup T \sim_c a_{2i}$'.

(We disregard the subscript on 'Θ' in the last line.)

Let us adopt, for the remainder of this section, a notation $x \sim' y$ to mean that there is a bijection $\pi : \bigcup x \longleftrightarrow \bigcup y$ such that $y = \{\pi``z : z \in x\}$. Notice that the pair $\{\sim_c, \sim'\}$ is a suite-of-congruence relations for the relation '$\bigcup T \sim_c a$'. AC implies that $\{\sim_c, \sim^{+'}\}$ too is a suite of congruence relations for "A is a partition and $\exists$ a partition $A' \sim^{+'} A \wedge b = \bigcup A'$". In fact, AC implies that—on partitions—$\sim^{+'}$ and $\sim'$ are the same relation, as follows. If $A = \{A_i : i \in I\}$ and $B = \{B_i : i \in I\}$ are disjoint families of sets with $|A_i| = |B_i|$ for all $i \in I$, then for $i \in I$ let C_i be the set of bijections between A_i and B_i. Use AC to pick a representative f_i from each C_i. Then $f = \bigcup_{i \in I} f_i$ is a bijection between $\bigcup_{i \in I} A_i$ and $\bigcup_{i \in I} B_i$. f is now a map in virtue of which $A \sim' B$.

Thus, since AC tells us that the size of $\bigcup B$ (when B is a partition) depends only on the size of B and on the sizes of its members, it enables us to represent the sum of a multiset of a single cardinal as a multiplication of that cardinal by the size of that multiset. This is the explanation for the otherwise rather mystifying name 'the multiplicative axiom' that Russell gave to the axiom of choice.[1] Russell wanted to regard multiplication as arising from sumsets of partitions all of whose elements are equipollent, rather than from cartesian products, and (as the *sutra* about the millionaire's footwear in Russell [1919] emphasises) AC is needed to show that the size of the sumset in these circumstances depends only on the size of the partition and the common size of its members. It might seem rather perverse to regard multiplication as arising from unions of disjoint families of sets all of one size rather than from cartesian product, given that it drags the axiom of choice into the picture, but there are natural cases where multiplication does arise in precisely this way. The simplest example is probably Lagrange's theorem that the order of every subgroup of a group divides the order of the group: if G' is a subgroup of G, the relation $x \sim_{G'} y$ iff $(\exists g \in G')(g \cdot x = y)$ is an equivalence relation, and all equivalence classes are the same size as G'.

The significance of AC in cardinal arithmetic is not just the fact that it is equivalent to an assertion of cardinal arithmetic ("all cardinals are comparable"), striking though that is; it's not even that AC in the form "products of nonempty sets are nonempty" implies that infinite products of nonzero cardinals don't turn out to be 0; what matters here is that because AC implies that $\{\sim_c, \sim^{+'}\}$ is a suite-of-congruences relations for "A is a partition and $\exists$ partition $A' \sim^{+'} A \ \wedge \ b = \bigcup A'$", it makes infinite sums and products *well-defined*. (The fact that it also makes them non-zero is a distraction that confused me for a long time.)

The case that Russell uses as an illustration can be used here to sharpen the distinction drawn in the last paragraph. We do not need the full strength of AC to deduce that all unions of countably infinite sets of disjoint pairs are the same size. This follows from (*inter alia*) the Prime ideal theorem ("PIT" for short) which is known to be weaker than the axiom of choice. PIT thereby has the consequence that $\{\sim_c, \sim^{+'}\}$ is a suite-of-congruence relations for "$\exists$ partition $A' \sim^{+'} A \ \wedge \ B = \bigcup A'$", so it is a consequence of PIT that this expression will belong to cardinal arithmetic as defined in the slogan. Notice that this is not the same as saying that PIT has consequences which belong to cardinal arithmetic. For example it does not yield the two other consequences of the axiom of choice just mentioned, which are assertions of cardinal arithmetic. It seems to be an open question whether or not PIT has any consequences which are assertions of cardinal arithmetic.

It would be perverse to pass $\sim'$ by without mentioning the context in which it most naturally belongs. We considered the suite of congruence

[1]The most accessible discussion of this by Russell is in Russell [1919] around p. 126.

relations $\{\sim_c, \sim'\}$ earlier, in connection with the relation $b = \bigcup A$. We wiggle b by $\sim_c$ and A by $\sim'$. Now if we wiggle A to $A' \sim A'$ this is because there is $f : \bigcup A \to \bigcup A'$. In general if we wiggle b at the same time, it won't be to $f``b$. We get an interesting notion of equivalence of pairs $\langle A, b\rangle$ with $\langle A', b'\rangle$ if we do not allow the two wigglings to be independent of each other in this way. We define this equivalence by saying $\langle A, b\rangle \sim \langle A', b'\rangle$ iff there is a permutation π of the universe such that both $\pi``b = b'$ and $\{\pi``x : x \in A\} = A'$. By this means we are defining an equivalence relation on tuples of things, rather than a tuple of equivalence relations on things.

4.2.3 *Ramsey Theory*

Parts of Ramsey theory definitely belong to cardinal arithmetic. Equipollence is a congruence relation for the ternary relation $R(A, B, I)$:

> For every $B' \sim_c B$ and for every partition of $\{X \subset B : X \sim_c I\}$ into two pieces there is a monochromatic set $A' \subseteq B'$ with $A' \sim_c A$

This means that expressions in the partition calculus of the form $\alpha \to (\beta)^\gamma_2$ belong to cardinal arithmetic construed virtually. (There is a problem with the subscript: the subscript cannot be a variable, though it can be any concrete natural number.) If we look at a four-place relation with an extra variable for a partition we find that the equivalence relation we need for the partition variable is not $\sim^+$ but the rather finer $\sim'$.

There is an extensive literature on this kind of equivalence, going back to Coret [1964], and it is of considerable interest. There are various reasons for not pursuing it further here. One is that it is not directly related to cardinal arithmetic. Another is that it does not give rise to clear criteria for individuation of virtual entities. A word is in order about this last point. Consider the following toy example. Let x be a set, and define $\sim$ on elements of $X = \mathcal{P}(x)$ by $y \sim z$ iff there is a permutation π of x such that $\pi``y = z$. This lifts to equivalence relations on tuples of elements of X, sets of elements of X and so on. The quotient is now a mess. There are $|x| + 1$ distinct elements of $X/\sim$, but there are many more than $\binom{|x|}{2}$ equivalence classes of doubleton subsets of X. If x had consisted of two elements a and b, X would be $\{\emptyset, \{a\}, \{b\}, \{a, b\}\}$ and $X/\sim$ will contain three equivalence classes, namely $\{\emptyset\}$, $\{\{a\}, \{b\}\}$ and $\{\{a, b\}\}$. But $\mathcal{P}(X)/\sim$ will not have *eight* equivalence classes but *twelve*. One equivalence class of the empty set: $\{\emptyset\}$; three equivalence classes of singletons: $\{\{\emptyset\}\}$, $\{\{\{a\}\}, \{\{b\}\}\}$ and $\{\{\{a, b\}\}\}$; four equivalence classes of doubletons: $\{\{\emptyset, \{a\}\}, \{\emptyset, \{b\}\}\}$, $\{\{\emptyset, \{a, b\}\}\}$, $\{\{\{a\}, \{b\}\}\}$, $\{\{\{a\}, \{a, b\}\}, \{\{b\}, \{a, b\}\}\}$; three equivalence classes of triples:$\{\{\{a\}, \{b\}, \{a, b\}\}\}$, $\{\{\emptyset, \{b\}, \{a, b\}\}, \{\emptyset, \{a\}, \{a, b\}\}\}$

and $\{\{\emptyset, \{a\}, \{b\}\}\}$; finally one equivalence class of a quadruple: $\{\{\emptyset, \{a\}, \{b\}, \{a, b\}\}\}$. The fact that $\{u', v'\}$ can fail to be equivalent to $\{u, v\}$ even when u is equivalent to u' and v is equivalent to v' falsifies extensionality and prevents us from thinking of the equivalence classes of tuples as virtual sets.

4.3 Virtual illfounded sets

So far we have been concerned with cardinal numbers, which arise from equivalence relations on sets. The long-term aim of this treatment is to deal with ordinal numbers, which arise from equivalence relations on *structures*. With a view to presenting illuminating examples in order of difficulty we will here have an interlude concerning equivalence relations—in fact isomorphisms—of various kinds of structures over sets rather than equivalence relations between sets. Recall from page 12 that $\mathfrak{X}$ is a structure with domain X. Lower case fraktur is not reserved in this way and the letter $\mathfrak{e}$ has a special use which will be explained below.

Strictly this notation gives rise to a typing discipline in the same way as pairing and unpairing did, and we should make this typing explicit. That point has already been made, and we will here revert to the habit, standard in modern mathematics, of assuming that this construct has been implemented set-theoretically in some way and not worrying about what it might be.

From our point of view these equivalence relations on structures are helpful because they raise some (but not all) of the complexities of the Burali-Forti paradox. In general they are interesting (indeed fashionable) because they provide simulations that give consistency proofs for various set theories with antifoundation axioms. It is also, perhaps, amusing to note that machinery erected to fake cardinals over sets can actually give nontrivial results by faking sets over sets! Instead of conservativeness results for theories in a new language, one gets relative consistency results for theories in the same language.

Our signature is to be $\in$ with a primitive pairing function as in chapter 3.3.1 with—in addition—a notation for structures: a overloaded tuple notation as mentioned above. We will use '$\mathcal{L}$' to denote this language, since we have been using this symbol to denote the language we are interpreting something into. '$\mathcal{L}^*$' will denote the language with—in addition to equality—the sole binary relation $\mathcal{E}$ ('$\mathcal{E}$' here is a pun on "edge" and "epsilon") whose variables will be upper-case $\mathcal{CALLIGRAPHIC}$ font letters, with the exception of $\mathcal{E}$ which is reserved for the pun.

4.3.1 *Irredundant trees*

This example is due to Dana Scott. (Scott never published it, as he says the idea had been discovered earlier by Specker, who never published it either).

In what follows a tree is not a partial order with knobs on but a digraph satisfying some special conditions.

(i) There is a unique vertex such that for all other vertices v there is a dipath (directed path) from it to v. This vertex is the **root**.

(ii) For all vertices v_1 and v_2 there is a unique vertex v_3 such that there are edge-disjoint dipaths from v_3 to v_1 and v_3 to v_2. In a tree $\langle V, E\rangle$ we say of two elements $v_1, v_2 \in V$ that v_2 is a **child** of v_1 if $v_1 E v_2$. This usage gives an obvious definition for the word **descendent**. Each element v of a tree gives rise to a subtree of that tree of which v is the root. If v is a child of the root we say that the subtree is an **child-tree**. A tree $\langle V, E\rangle$ is **irredundant** if it has trivial automorphism group.

Let us write $\langle V, E\rangle \mathfrak{e} \langle V', E'\rangle$ if there is a vertex $v' \in V'$ such that (i) v' is a child of the root of $\langle V', E'\rangle$, (ii) $\langle V, E\rangle \simeq$ the restriction of $\langle V', E'\rangle$ to the descendents of v'. This is a binary relation between irredundant trees for which tree-isomorphism is a congruence relation.

Now we define $\mathcal{I}$ in the standard way so that $\mathcal{I}$ of $\mathcal{X}\mathcal{E}\mathcal{Y}$ is $\mathfrak{X} \mathfrak{e} \mathfrak{Y}$ (so that $\mathcal{I}$ sends calligraphic font variables to variables ranging only over irredundant trees) and $\mathcal{I}$ of '=' is the symbol for tree isomorphism.

This naturally gives rise to a relative consistency proof for some kind of set theory, in the sense that whenever we start with a theory S of $\mathcal{L}$ the set $\{\phi \in \mathcal{L}^* : S \vdash \mathcal{I}(\phi)\}$ is a theory in the language of set theory which is consistent relative to S. What does this theory contain?

We start by proving that this theory contains extensionality. Extensionality in $\mathcal{L}^*$ is

$$(\forall \mathcal{X})(\forall \mathcal{Y})(\mathcal{X} = \mathcal{Y} \longleftrightarrow (\forall \mathcal{Z})(\mathcal{Z}\mathcal{E}\mathcal{X} \longleftrightarrow \mathcal{Z}\mathcal{E}\mathcal{Y}))$$

and $\mathcal{I}$ of this is

$$(\forall \mathfrak{X})(\forall \mathfrak{Y})(\mathfrak{X} \sim \mathfrak{Y} \longleftrightarrow (\forall \mathfrak{Z})(\mathfrak{Z} \mathfrak{e} \mathfrak{X} \longleftrightarrow \mathfrak{Z} \mathfrak{e} \mathfrak{Y}))$$

The left-to-right direction is easy. It is for the other direction that we need the irredundancy condition. Suppose $(\forall \mathfrak{Z})(\mathfrak{Z} \mathfrak{e} \mathfrak{X} \longleftrightarrow \mathfrak{Z} \mathfrak{e} \mathfrak{Y})$. Every child-tree of $\mathfrak{X}$ is isomorphic to an child-tree of $\mathfrak{Y}$, and indeed a unique child-tree, since, by irredundancy there can be only one. Naturally every child-tree of $\mathfrak{Y}$ is similarly isomorphic to a unique child-tree of $\mathfrak{X}$. We can actually say slightly more than this. Not only is each child-tree $\mathfrak{Y}'$ of $\mathfrak{Y}$ isomorphic to a unique child-tree of $\mathfrak{X}$ but this isomorphism is unique: if there were more than one isomorphism between them $\mathfrak{Y}'$ would have an automorphism, contradicting irredundancy of $\mathfrak{Y}$. So the isomorphism we want is simply the union of all the (uniquely identifiable) isomorphisms for all the child-trees of $\mathfrak{Y}$ and of $\mathfrak{X}$.

Verifying that the axiom of power set is in $\{\phi \in \mathcal{L}^* : S \vdash \mathcal{I}(\phi)\}$ relies on some rudimentary comprehension axioms. Given a tree $\mathfrak{X}$ one can form lots of fragments of it obtained by deleting some children of the root along with all their descendents. These fragments are irredundant too of course. Disjoint copies of all these trees can be made and all put together into a new tree by making the root of each the child of new root. This is a tree all right, and we must check that it is irredundant. Since any child-tree of this tree is irredundant (it is one of the fragments alluded to above) the only way the new tree can fail to be irredundant is if two of its child-trees are isomorphic. But if there were two such child-trees, an isomorphism between them would at the very least extend an isomorphism between two child-trees of the original tree. Such an isomorphism could be extended to an automorphism of $\mathfrak{X}$ merely by fixing everything not in those two child-trees. This contradicts irredundancy of $\mathfrak{X}$. The $\mathcal{L}^*$-version axiom of power set holds in virtue of the existence of this tree for all $\mathfrak{X}$.

Similarly an axiom scheme of separation will hold for $\mathcal{E}$ as long as separation holds in the set theory in which we are developing the theory of trees. Pairing too is easy to verify.

A subtlety arises with the remaining axiom of set theory. Consider the axiom of sumset. Tackling the problem by analogy with the way we dealt with power set what one naturally wants to do on being presented with a tree $\mathfrak{X}$ is find a tree whose child-trees are precisely those that are isomorphic to child-trees of child-trees of $\mathfrak{X}$. How are we to do this? Well, take all the grandchildren of the root of $\mathfrak{X}$, disjoin all the subtrees they give rise to, and stick them all together with a new root as we did in the case of the power set axiom. The trouble with this is that this collection may contain lots of isomorphic trees and we have to throw away all but one. The obvious way to do this is to pick one from each isomorphism class by means of the axiom of choice. One might expect that in order to verify an axiom for $\mathcal{E}$ it should be sufficient to check that that axiom held in the set theory in which we were developing our theory of trees to start with (as was the case for power set, pairing and separation). The need for the axiom of choice here is most displeasing and unwelcome. Later we will see a way of getting round it, but it involves implementations as well as simulation. The path to it leads through iterated virtuality, to which we now turn.

Chapter 5

Iterated Virtuality in Cardinal Arithmetic

What we saw in section 3.3 would be expressed by a modern computer scientist in language somewhat like the following. Given a type α, then as long as it's a type whose internal structure makes it a type of *sets* in some suitable sense, we can invent a type `cardinal-of` α. Also, for any such type α we can invent a type `set-of` α and `set-of-set-of` α and even `set-of-set-of-cardinal` α and indeed `set-of-set-of-set-of-cardinal` α, and so on, all in the same style. All of these types are virtual types "over" (should one feel the need for a preposition) the type α. They are as it were *singly virtual*. What we have is a higher-order theory of singly virtual (unimplemented) cardinals.

5.1 Doubly virtual cardinals

"Singly" virtual? The higher-order language for cardinal arithmetic that we developed in the last chapter is a language that has in it syntactic gadgetry for all predicates for which the equivalence relations $\sim_c \ldots$ formed a suite of congruence relations. In particular it has shadows of the membership relation $\in$ of the original set theory—or rather it has one shadow at each level. This makes it formally a language for set theory, and this has the consequence that the development that gave us cardinal arithmetic from set theory can be repeated to give us a theory of cardinals of sets of cardinals. These cardinals will be *doubly* virtual. We will certainly have to deal with an iteration of this kind when analysing the Burali-Forti paradox so it may be an idea to have a look at iteration when it first comes up, in the slightly more straightforward theatre of cardinal arithmetic.

To simulate cardinals of sets of cardinals we need to be able to simulate injections from sets of cardinals into sets of cardinals. We have already seen how to simulate sets of cardinals but not yet injections from sets of cardinals to sets of cardinals and to do this we can use Henkin quantifiers as we did in the first instance.

If we think of bijections in terms of ordered pairs then we would need to simulate ordered pairs of cardinals and sets of ordered pairs of cardinals. So far we have used Roman letters for variables over sets, and Greek letter for variables over cardinals. We use calligraphic font for ordered pairs of cardinals because we need something that is neither Greek nor Roman. Ordered pairs of cardinals we deal with by adding to $\mathcal{L}^*$ a three-place relation $\texttt{cardinal-pair}(\alpha, \beta, \mathcal{P})$ whose intended meaning is that $\mathcal{P}$ is the ordered pair of α and β, and from this we will define a singular term with free variables in it to behave like a function letter for pairing of cardinals. (All this is as in section 2.2.)

Now from what formula in set theory could this relation arise? The obvious candidate is $(\exists a' \sim_c a)(\exists b' \sim_c b)(p = \langle a', b' \rangle)$, and we will need an equivalence relation R on pairs such that $\{\sim_c, \sim_c, R\}$ is a suite of congruence relations for `cardinal-pair`. Clearly this equivalence relation must be that which holds between two pairs $\langle x_1 x_2 \rangle$ and $\langle y_1, y_2 \rangle$ iff $x_1 \sim_c y_1$ and $x_2 \sim_c y_2$, and we shall write it as '$x \sim_{cp} y$'. Clearly the appropriate equivalence relation to be $\mathcal{I}$ of '=' between variables ranging over sets of ordered pairs of cardinals will be '$\sim^+_{cp}$'.

Then we could extend definition 4.1 by adding the clauses

$\mathcal{I}$ of $\texttt{cardinal-pair}(\alpha, \beta, \mathcal{P})$ is $(\exists a' \sim_c a)(\exists b' \sim_c b)(\mathcal{P} = \langle a', b' \rangle)$.

$\mathcal{I}$ of variables ranging over ordered pairs of cardinals shall be the corresponding lower-case Roman variable.

$\mathcal{I}$ of '=' between singular terms denoting ordered pairs of cardinals shall be $\sim_{cp}$;

$\mathcal{I}$ of variables ranging over sets of ordered pairs of cardinals shall be the corresponding upper-case Roman variable.

$\mathcal{I}$ of '=' between terms denoting sets of ordered pairs of cardinals shall be $\sim^+_{cp}$'.

... between clauses 6 and 7 of that definition, leaving the old catch-all clause 7 at the end.

This will ensure that the pairing and unpairing axioms are satisfied. (e.g. $\mathcal{I}$ of $\texttt{cardinal-pair}(\alpha, \beta, \zeta) \wedge \texttt{cardinal-pair}(\alpha, \beta, \zeta') \rightarrow \zeta = \zeta'$ will be a theorem of T). Now we have a notation for ordered pairs of cardinals, since the singular term in the new language which denotes the ordered pair of α and β is of course '$(\iota Z)\texttt{cardinal-pair}(\alpha, \beta, Z)$'. Then the analogue of formula 4.1 that we want is a formula stating that the sets X and Y stand in the relation from which equipollence of sets of cardinals will arise, namely that there is a set $\subseteq X \times Y$ that relates everything in X to something in Y and *vice versa*, and whenever x_1 and x_2 are two members of X there is a thing in Y related to them both iff $x_1 \sim_c x_2$ and whenever y_1 and y_2

are two members of Y there is a thing in X they are both related to iff $y_1 \sim_c y_2$.

$$(\exists F \subseteq X \times Y)(\forall x \in X)(\exists y \in Y)(\langle x, y\rangle \in F) \wedge (\forall y \in Y)(\exists x \in X)(\langle x, y\rangle \in F) \wedge (\forall x_1 x_2 \in X)((\exists y \in Y)(\langle x_1, y\rangle \in F \wedge \langle x_2, y\rangle \in F) \longleftrightarrow x_1 \sim_c x_2) \wedge (\forall y_1 y_2 \in Y)((\exists x \in X)(\langle x, y_1\rangle \in F \wedge \langle x, y_2\rangle \in F) \longleftrightarrow y_1 \sim_c y_2)$$

Once we have simulated all these things we will have a language that has pairs-of and sets-of and sets-of-pairs-of over cardinals as well as sets. The point is that the language will then have all the apparatus that $\mathcal{L}$ had when we showed how to simulate (singly virtual) cardinals in it. Then we will have all the apparatus we need to repeat the process that got us to that stage.

But this is incredibly unwieldy, and we will continue to use Henkin quantifiers as before, so that we write '$X \sim_{c\dagger} Y$' for

$$\begin{matrix}(\forall x \in X)(\exists y \in Y) \\ (\forall y' \in Y)(\exists x' \in X)\end{matrix} (x \sim_c x' \longleftrightarrow y \sim_c y')$$

Armed with these we can develop a theory of (doubly virtual) cardinals of sets of (singly virtual) cardinals just as we developed a theory of (singly virtual) cardinals of sets above.

To develop the higher (but still singly virtual) types alluded to at the outset of this chapter we will also need to know what $\mathcal{I}$ of '$x = \langle y, z\rangle$' is when 'x', 'y' and 'z' are variables ranging over $(\text{sets})^{n+1}$ of cardinals of sets. (as it were $n + 1$th order Greek variables.) We do this by recursion on n. The case $n = 0$ we have just disposed of, and the obvious device for dealing with the recursion case is cartesian product, so that $\mathcal{I}$ of the formula that says that x, y and z are $(\text{sets})^{n+1}$ of cardinals of sets and z is the pair of x and y is the result of sticking $\mathcal{I}$ of the each variable (x, y and z) in the appropriate position in '$x = y \times z$'.

Having extended definition 4.1 by adding the clauses on page 54 we now have an interpretation into $\mathcal{L}$ of all the syntactic gadgets that enables us to simulate cardinals in $\mathcal{L}$ in the first place. This means the process can now be repeated.

5.2 Multiply virtual cardinals

There was a young Lady of Spain
Who was terribly sick on a train
Not once but again
and again and again
and again and again and again

Of course having done it twice we can do it thrice, and so on. This will show that starting with T a theory of sets, we can develop a many-sorted theory T^∞ in a language $\mathcal{L}^\infty$ where the sorts are to be found in the following collection: V, `cardinal-of` V, `cardinal-of-set-of-cardinal-of-`V, `cardinal-of-set-of-cardinal-of-set-of-cardinal-of` V, ... At each stage, where we progress from a sort α to a sort `cardinal-of-set-of` α we go through a manœuvre just like the one in section 3.2.

This is a lot more complicated than the ωth-order theory of section 3.2 since it involves iterating that procedure ω times. However for it too we can develop a canonical simulation.

To explain how to do this we have to cleave hard unto the ideas of *singly* virtual, *doubly* virtual and so on. Notice that every formula of $\mathcal{L}^\infty$ has an upper bound on the degree of virtuality of the entities of which it speaks. At each stage when we erect a new language for describing entities virtual over the entities of the old language we have a canonical simulation as in definition 3.2, and we can compose this on the right with the preëxisting canonical simulation to obtain a new one of higher degree. Thus for every formula ϕ of $\mathcal{L}^\infty$ there is a minimal n such that it is in the domain of a canonical simulation of degree n. The global canonical simulation for $\mathcal{L}^\infty$ simply sends ϕ to whatever it is sent to by the canonical simulation of degree n.

5.2.1 *The parts that virtualism cannot reach*

So we now have a virtualist theory of cardinals: cardinals as virtual enitites over sets, doubly virtual cardinals as virtual entities over sets of cardinals, and so on. Does this provide a semantics for all the expressions that we thought belonged to cardinal arithmetic?

It covers most of what is normally regarded as cardinal arithmetic. Indeed it even covers some assertions which are perhaps not always thought of as part of cardinal arithmetic, such as some versions of Ramsey's theorem, as we have seen. However there are several phenomena that it does not cover which we should consider here, and this was spotted years ago by

Frege (1903) (in the appendix to volume II, bottom of p.255).[1]

5.2.1.1 *The Paris-Harrington theorem*

The first example is the Paris-Harrington concept of *relative largeness* of finite sets of integers: a set is relatively large if its cardinality is larger than its smallest element. If we try to type '$|X| \geq min(X)$' we find that X is of type `set-of cardinal-of` α but $|X|$, which should be of type `cardinal-of` α, must of course have type `cardinal-of set-of cardinal-of set-of` α and these two types will not unify.

It may not be too fanciful to make a connection between the great strength of the Paris-Harrington theorem and its type-violating nature.

5.2.1.2 *Exponentiation*

Exponentiation is problematic, because if we think of it as iterated multiplication the exponent is doubly virtual and the base is singly virtual. On the face of it this ought to mean that we cannot apply the operation to two numbers of the same flavour, which would be disasterous. We can get round this by insisting that exponentiation arises from the set-of-all-functions construction, and it can indeed be seen as arising in this way. But often it doesn't. Let $\phi(n)$ be the number of natural numbers less than n that are prime to n. This is *Euler*'s totient function. Euler's theorem states:

If $m \geq 2$ and a, m coprime, $a^{\phi(m)} \equiv 1 \pmod{m}$

The usual proof goes as follows. Let U be the set of naturals $n < m$ prime to m. There are $\phi(m)$ of them. Let Ua be the result of multiplying all the things in U by a. If we multiply a thing in U by a (mod m) we just get another thing in U because a and m are coprime. For the same reason we never get the same thing twice, so multiplying by a just permutes U. So

$$(a \cdot u_1) \cdot \ldots (a \cdot u_{\phi(m)}) = u_1 \cdot \ldots u_{\phi(m)}$$

On the left collect all the a's together to get

$$a^{\phi(m)} \cdot u_1 \cdot \ldots u_{\phi(m)} = u_1 \cdot \ldots u_{\phi(m)}$$

and we can divide both sides by $u_1 \cdot \ldots u_{\phi(m)}$ since it is prime to m (being a product of things prime to m) to get

[1] Thanks are due to Dr. Michael Potter for drawing this to my attention.

$$a^{\phi(m)} = 1(\text{mod } m)$$

■

This can be contorted into an assertion purely about sets and not about sets of numbers. $\sim_c$ is a congruence relation for '$(\forall abc)((a \times b \sim_c x \wedge a \times c \sim y) \to (\exists z)(a = \{z\}))$' after all, but the usual proof definitely uses doubly virtual numbers. Presumably there is a proof using singly virtual cardinals only but we shouldn't be surprised if all such proofs are much nastier than Euler's: extending a language will often give new proofs of old facts.

There is another elementary consequence of the axiom of choice, which is normally regarded as part of cardinal arithmetic, and yet does not have a virtualist account. This is the assertion that a union of α disjoint sets each of size β is equipollent to a union of β disjoint sets each of size α. We encountered half of this problem earlier, when we introduced multisets of cardinals and showed how to use the axiom of choice to show that Σ ("sums") is a well-defined operation taking multisets of cardinals to cardinals. Σ was defined there as an operation taking multisets of cardinals as its argument, and we didn't consider the size of the multiset to be an argument. If we wish to assert that "a union of α disjoint sets each of size β is equipollent to a union of β disjoint sets each of size α" then we not only need enough AC to make the sums well-defined but we also notice that in the text just quoted the first occurrence of α' is doubly virtual and the second occurrence is singly virtual, and *vice versa* for the occurrences of 'β'. I drew attention earlier (p. 47) to Russell's use of the axiom of choice to ensure that these sums are well-defined, and to this being his reason for calling AC the 'multiplicative axiom'. However for this to work he also needs to be able to disregard the difference between singly and doubly virtual cardinals.

We will return to this tangle after we have borrowed some apparatus from ordinal arithmetic.

5.3 Untyped invariant arithmetic

There is an obvious way to loosen the virtualist analysis to include concepts like 'relatively large' within the ambit of cardinal arithmetic. It is obvious that we want to: the analysis that excludes $3 \in 5$ was never meant to exclude relative largeness. Fortunately a remedy appears to be at hand. It's pretty obvious that whether or not a set of cardinals is relatively large will not depend on how cardinals are implemented. So the fallback position is as follows. Define what an implementation of cardinal arithmetic is, and then say that a formula of the original language belongs to cardinal arithmetic iff its truth-value is not affected by changing the implementation. That will let in relative largeness but still exclude $3 \in 5$.

Very well, we add to our original language $\mathcal{L}$ for set theory a new function letter, 'f', say, with axioms to say that f is a function and an implementation of cardinal arithmetic, that is $(\forall yx)(f(x) = f(y) \longleftrightarrow x \sim_c y)$.

For this to be a sensible thing to do, we have to ensure that enlarging the language in this way does not give us any new comprehension axioms. Of course if f is definable then there will be no new set existence theorems, but we want something more general. We will need a typing system.

Accordingly for the rest of this chapter let $\mathcal{L}_\in$ be the language of pure set theory and let $\mathcal{L}^\dagger$ be the language of set theory with primitive cardinals ("$x = |y|$"). Structures for $\mathcal{L}^\dagger$ accordingly have a domain, and two binary relations: membership and is-cardinal-of.

We have a notion of *degree* for formulæ of this language which is precisely analogous to the sense introduced for $\mathcal{L}^\infty$ at the beginning of section 5.2. We will also need a notion of typing,

Definition 5.1

(1) One constant: `set`;

(2) Two operations: `set-of` and `cardinal-of`. `cardinal-of` cannot be applied to values of the operation `cardinal-of`. (Only sets have cardinals!)

(3) A function σ from variables in (a formula ϕ) to elements of the type algebra is a **typing** when it satisfies the three conditions:
(i) If '$x \in y$' occurs in ϕ and σ of 'x' is the type `set` then σ of 'y' is the type `set`.
(ii) If '$x \in y$' occurs in ϕ and σ of 'x' is the type α (with $\alpha \neq$ `set`) then σ of 'y' is the type `set-of` α.
(iii) If '$y = |x|$' occurs in ϕ and σ of 'x' is the type α then σ of 'y' is the type `cardinal-of` α.

Finally a formula is **typed** if there is a typing defined on all the variables in it.

We now restrict our comprehension scheme to formulæ in $\mathcal{L}^\dagger$ that are typed in this sense. The point of this typing is that we will be able to show that every formula that is typed is equivalent to one that has no occurrences of f in it.

This will ensure that the obvious extension of a theory in $\mathcal{L}_\in$ to a theory in $\mathcal{L}^\dagger$ is conservative.

5.4 Implementation-insensitivity

The time has now come to return to the idea of implementation-independence first raised on p. 14.

Definition 5.2

Let us say that a formula $\phi \in \mathcal{L}^\dagger$ is **implementation-insensitive** for T if whenever $\mathfrak{M}$ is a model for T and $\mathfrak{M}_1^\dagger$ and $\mathfrak{M}_2^\dagger$ are expansions of $\mathfrak{M}$ obtained by adding a relation that is the interpretation of the cardinal-of relation symbol (so that $\mathfrak{M}_1^\dagger$ and $\mathfrak{M}_2^\dagger$ are $\mathcal{L}^\dagger$-structures) then

$$\mathfrak{M}_1^\dagger \models \phi \longleftrightarrow \mathfrak{M}_2^\dagger \models \phi$$

Now since a different choice for a function that is the interpretation of the cardinal-of function symbol can be obtained simply by composing one such function on the left with a permutation of the domain, we might expect there to be a formulation of implementation-independence in terms of permutations.

We need the notion of a *permutation model* of an $\mathcal{L}^\dagger$ theory. There are two ways of approaching this idea.

Definition 5.3 If $\mathfrak{M} = \langle M, \in, c\rangle$ is a $\mathcal{L}^\dagger$-structure and σ is a permutation of M then $\mathfrak{M}^\sigma = \langle M, \in, c_\sigma\rangle$ (where $c_\sigma(x, y)$ iff $c(\sigma(x), y)$ so that '$c(\sigma(x), y)$' is read as '$\sigma(x) = |y|$') is another $\mathcal{L}^\dagger$-structure and is a **permutation model** of $\mathfrak{M}$.

If ϕ is an $\mathcal{L}^\dagger$-formula then ϕ^σ is the result of replacing in ϕ all occurrences of '$y = |x|$' by '$\sigma(y) = |x|$'.

The connection between the two is that if $\mathfrak{M} = \langle M, \in, c\rangle$ is a $\mathcal{L}^\dagger$-structure and σ is a permutation of M and $\phi \in \mathcal{L}^\dagger$ then $\mathfrak{M}^\sigma \models \phi$ iff $\mathfrak{M} \models \phi^\sigma$.

Definition 5.4 If $(\forall\mathfrak{M})(\mathfrak{M} \models \phi \longleftrightarrow (\forall\mathfrak{M}')(\mathfrak{M}'$ is a permutation model of $\mathfrak{M} \rightarrow \mathfrak{M}' \models \phi))$ we say ϕ is **invariant**.

We will see that implementation-insensitivity implies invariance which implies typed but that the arrows cannot be reversed.

The motivation for this device of permutation model is that if $\mathfrak{M}$ is a model of set theory with implemented cardinal numbers and σ is a permutation then $\mathfrak{M}^\sigma$ differs from $\mathfrak{M}$ only in having a different implementation of cardinals: it should satisfy the same assertions of cardinal arithmetic. This is theorem 5.1.

The converse is not true. There are circumstances in which we can expand $\mathfrak{M}$ into $\mathfrak{M}_1^\dagger$ and $\mathfrak{M}_2^\dagger$ which are not permutation models of each other. Suppose $\mathfrak{M}$ is a model of GB with global choice. This theory proves that all proper classes are the same size. Since the collection of cardinals cannot be a set, it must be implemented as a proper class, and since all proper classes are the same size, it can be implemented as any proper class, in particular as V. But clearly no $\mathfrak{M}_1^\dagger$ that makes every set a cardinal can be a permutation model of a $\mathfrak{M}_2^\dagger$ in which only von Neumann ordinals can be cardinals.

Definition 5.5

(1) If σ is a permutation, let $j(\sigma)$ be $\lambda x.\sigma``x$.
(2) If σ is a permutation of M with the property that $j^n(\sigma)$ is defined for all $n \in \mathbb{N}$ we say that σ is **setlike**.
(3) If σ is a permutation of an $\mathcal{L}^\dagger$ structure $\mathfrak{M}$ let $K(\sigma)$ be the function sending $|x|$ to $|\sigma(x)|$.

The third part of definition 5.5 looks a bit odd. If σ is an arbitrary permutation then this definition doesn't make sense, since σ could send two things the same size to two things of different sizes. On the other hand if σ is j of something (and fortunately it always will be in cases of interest!) how can $K(\sigma)$ be anything other than the identity? Surely $|x| = |\sigma``x|$ always? But beware: $\sigma$ being setlike guarantees that $\sigma``x$ is a set as long as x is, and it guarantees that x and $\sigma``x$ are *externally* the same size but it doesn't promise that the graph of σ restricted to x is a set of the model, which is what would be needed for them to be *internally* the same size.

The notation K is a pun on 'cardinal' and the K combinator. This is because if σ is a permutation of an $\mathcal{L}^\dagger$ structure $\mathfrak{M}$ that is a model of the class form of replacement ("the image of a set in a class is a set") then $K(j(\sigma))$ is simply the identity, so that K is K (the combinator K) of the identity.

Suppose σ is a permutation of an $\mathcal{L}^\dagger$ structure $\mathfrak{M}$ that is *not* a model of the class form of replacement so that we cannot be sure that w and $\sigma``w$ are the same size according to $\mathfrak{M}$. Can we be sure that if $w$ and $w'$ are the same size according to $\mathfrak{M}$ then $\sigma``w$ and $\sigma``w'$ are the same size according to $\mathfrak{M}$? We need this to know that $K(\sigma)$ is defined. But this is easy. Translation by $j(\sigma)$ turns any bijection between w and w' into a bijection between $j(\sigma)(w')$ and $(j(\sigma))(w)$.

We've defined $K(j(\sigma))$ only on the things that are (implemented) cardinals of $\mathfrak{M}$, but it can be extended naturally to a permutation of M, because $K(j(\sigma))$ permutes those cardinals: we can just fix everything else (as before).

Theorem 5.1 *As long as ϕ is typed in the sense of definition 5.1 and σ is setlike then $\mathfrak{M} \models \phi$ iff $\mathfrak{M}^\sigma \models \phi$, which is to say iff $\mathfrak{M} \models \phi^\sigma$.*

Proof: The trick is to show that, as long as ϕ is typed in the sense of definition 5.1, then ϕ and ϕ^σ are interdeducible, because we can eliminate all occurrences of 'σ' from ϕ^σ and obtain ϕ.

Let ϕ be typed in the sense of definition 5.1 and fix a typing of it: we will speak of the type assigned to each variable by the fixed typing as the **type** of that variable.

Prima facie 'ϕ^σ' is liable to have some variables within it which (on some of their occurrences) are prefixed by 'σ' and at others are not. If we can manipulate ϕ^σ into a form where, for each bound variable, all occurrences of that variable have the same prefix, then we we can simply delete that prefix.

If ϕ is typed then we can do this, and—once we have settled on a typing for ϕ—we can use it to determine, for each variable 'x' in ϕ, the uniform prefix that all occurrences of 'x' will have. This we do by recursion on the words of the type algebra.

A variable given type `set` has no prefix;
A variable given type `cardinal-of set` has prefix σ;
If 'x' has prefix τ and '$x \in y$' appears in ϕ then 'y' has prefix $j(\tau)$;
If 'x' has prefix τ and '$y = |x|$' appears in ϕ then 'y' has prefix $K(\tau)$.

Let us see how this works and why. Suppose the atomic subformulæ in ϕ are '$x \in y$', '$z = |y|$', '$z \in w$', '$u = |w|$' and '$u \in v$'. There is a typing giving to 'x' and to 'y' the type `set`, to 'z' the type `cardinal-of set`, to 'w' the type `set-of cardinal-of set`, to 'u' the type `cardinal-of set-of cardinal-of set`, and finally to 'v' the type `set-of cardinal-of set-of cardinal-of set`.

These atomic subformulæ reappear in ϕ^σ as '$x \in y$', '$\sigma(z) = |y|$', '$z \in w$', '$\sigma(u) = |w|$' and '$u \in v$'.

Now, by definition of j, we have $z \in w \longleftrightarrow \sigma(z) \in (j(\sigma))(w)$ so by substitutivity of the biconditional we can replace all occurrences of '$z \in w$' by '$\sigma(z) \in (j(\sigma))(w)$'. This has the consequence that all occurrences of 'z' have the same prefix, and that it is 'σ' which is the prefix appropriate to the type of 'z'.

By definition 5.5 in general '$a = |b|$' can be replaced by '$(K(\sigma))(a) = |\sigma(b)|$' and in particular '$\sigma(u) = |w|$' can be replaced by '$K(j(\sigma)) \cdot \sigma(u) = |(j(\sigma))(w)|$'.

This has the consequence that all occurrences of 'w' have the same prefix, and that it is '$j(\sigma)$' which is the prefix appropriate to the type of 'w'.

The variable 'u' now appears in two (rewritten) subformulæ of ϕ^σ, namely '$K(j(\sigma)) \cdot \sigma(u) = |(j(\sigma))(w)|$' and '$u \in v$'. Now we just need j, because $u \in v \longleftrightarrow K(j(\sigma)) \cdot \sigma(u) \in (j(K(j(\sigma)) \cdot \sigma))(v)$. Once we have done this, all occurrences of 'u' have the same prefix, namely '$K(j(\sigma)) \cdot \sigma$' Similarly all occurrences of 'v' end up with a prefix '$j(K(j(\sigma)) \cdot \sigma)$'.

Once all occurences of a variable have the same prefix, then the prefix can be deleted from the variable as soon as the variable is bound, since the prefix always denotes a permutation, and if σ is a permutation then $(\forall x)(\ldots \sigma(x) \ldots)$ can be rewritten as $(\forall x)(\ldots x \ldots)$. ∎

Clearly the Paris-Harrington formula is not well-typed according to the scheme of definition 5.1, so we cannot expect to prove it invariant by appealing to theorem 5.1. However, the following train of thought leads one to expect that it ought to be invariant nevertheless.

A formula ϕ is implementation-independent if ϕ and ϕ^σ are interdeducible, where ϕ^σ is the result of replacing all occurrences of '$y = |x|$' by '$\sigma(y) = |x|$', or—once we have introduced a singular term '$|x|$' as above, by replacing all occurrences of '$|x|$' in ϕ by $\sigma(|x|)$,

Let us take the formula 'x is relatively large' to illustrate what happens.

$(x \text{ is relatively large})^\sigma$

is

$$(\forall y)((\forall z)(z = |y| \rightarrow z \in x) \rightarrow x \hookrightarrow_c y)^\sigma$$

apply σ

$$(\forall y)((\forall z)(\sigma(z) = |y| \rightarrow z \in x) \rightarrow x \hookrightarrow_c y)$$

Now $z \in x \longleftrightarrow \sigma(z) \in \sigma``x$ and $x \hookrightarrow_c y \longleftrightarrow \sigma``x \hookrightarrow_c y$ whence

$$(\forall y)((\forall z)(\sigma(z) = |y| \rightarrow \sigma(z) \in \sigma``x) \rightarrow \sigma``x \hookrightarrow_c y)$$

Now reletter $\sigma(z)$ as z getting

$$(\forall y)((\forall z)(z = |y| \rightarrow z \in \sigma``x) \rightarrow \sigma``x \hookrightarrow_c y)$$

which is simply the assertion that $\sigma``x$ is relatively large.

Beware! This manipulation depended on the innocent-looking assumption that $x \hookrightarrow_c y$ if and only if $\sigma``x \hookrightarrow_c y$, which of course follows from $x$ and $\sigma``x$ being equipollent. This is true as long as the restriction of σ to a set is also a set, but the assumption that the restriction of a class to a set is a set is a form of replacement and not all set theories have it. If σ is merely setlike, then $\sigma``x$ will be a set whenever x is, but it might not be the same size as x. Sentences like the Paris-Harrington theorem are invariant as long as we consider invariance under permutations that are sets, or locally are sets in the manner that definable functions are almost sets in ZF (images of sets in them are sets, and their restrictions to sets are sets).

There are three confounding factors.

(1) The size of the range of the cardinal-of function. This is what prevents implementation-insensitivity from being the same as invariance. (See page 60).
(2) $(\forall\sigma)(\phi \longleftrightarrow \phi^\sigma)$ becomes easier to prove if there are fewer σ. Only considering σ that are sets (as opposed to setlike) allows some untyped things to be invariant.
(3) $(\forall\sigma)(\phi \longleftrightarrow \phi^\sigma)$ also becomes easier to prove if we add more axioms!

Were it not for (1) we could deal with (2) and (3) by restricting our attention to a suitably minimalist theory (to take care of (3)) and let the quantifier in (2) range over all setlike permutations. Then we would be able to prove a completeness theorem for well-typed formulæ to the effect that a formula is well-typed iff it is implementation-insensitive. (This would parallel the completeness theorem for Rieger-Bernays permutation models and stratified formulæ in Forster [1993].) But (1) makes that impossible. It would be possible to show by standard methods that for a suitable weak theory the invariant sentences are precisely the well-typed ones. The effect of such a theorem would be to drive a wedge between well-typed formulæ and formulæ like '$|x| > min(x)$' rather than between '$|x| > min(x)$' and '$3 \in 5$'. But these last two are separated by implementation-insensitivity, the first being implementation-insensitive and the second not. That draws the line we want.

We can capture syntactically the idea of invariance as follows. Set up a typing system in the style of definition 5.1 with two types, `cardinal` and `set`. In subformulæ like '$x = |y|$', 'x' must be of type `cardinal` and 'y' must be of type `set`; in subformulæ like '$x \in y$', 'y' must be of type `set`; finally in subformulæ like '$x = y$', 'x' and 'y' must be of the same type.

We need a concept of permutation model for theories of this signature. For ϕ an expression in this language, and σ a permutation of the carrier set, ϕ^σ is the result of replacing every variable 'x' of type `cardinal` with '$\sigma(x)$'. Then we can prove that every formula typed according to this scheme is invariant as follows. However, to do this we need some nontrivial set-theoretic assumptions. This is because we want σ"x to have the same cardinal as x whenever σ is a setlike permutation. (Remember: the Paris-Harrington formula will come out as typed according to this plan, and we want it to be implementation-insensitive as well. But for it to be implementation-insensitive it must be impossible for x and σ"x to be given different cardinals, and this requires the enveloping theory to prove that there is a map between x and σ"x. This map does not have to be σ restricted to f, but a map there must be. In particular this will mean that there will always be a map between x and $\{\{y\} : y \in x\}$. Whereas this is provable in ZF and all its usual subsystems, it is not provable in the Quine systems.

Finally, let us note that Gauntt *op. cit.* showed that it is consistent—even with ZF—that there should be no implementation of cardinal arithmetic at all. The von Neumann implementation, which is faithful (well, nearly faithful: see p 33), and sends every set to the least von Neumann ordinal equipollent to it, is not available if the axiom of choice fails, and Scott's trick (see p. 85) is unavailable if not every set is the same size as a wellfounded set.

5.5 Iterated virtuality and reflection

In section 4.3.1 we considered the prospect of an interpretation arising from an isomorphism relation on irredundant trees, which have a natural embedding relation between them that has some of the formal properties of set membership. We discovered that the obvious way of proving that the axiom of sumset is sent by the canonical simulation to a provable formula of the original theory involved having the axiom of choice in the original theory. It was said there that there is a way of avoiding the axiom of choice that exploits iterated virtuality. We are now in a position to discuss this possibility.

Irredundant trees are merely one way of faking set theory over set theory. There are also acyclic pointed graphs, pointed wellfounded extensional relations, among others. The feature common to all these devices is that they all look like codings of the membership relation restricted to the transitive closure of a set: one can think of these structures as different flavours of *pseudoset*. Each flavour of pseudoset admits an embedding relation $\mathcal{E}$ (between pseudosets) which looks like set membership, and therefore like the structural relation within a pseudoset. In these circumstances we have a precise logical analogue of the T function that we encountered above with ordinals. T of a relational type $\mathcal{A}$ is the relational type of the structure $\langle\{\mathcal{B} : \mathcal{B}\mathcal{E}\mathcal{A}\}, \mathcal{E}\rangle$. (or at least something like that: the detail will be quite sensitive to the flavour of pseudoset being used.) This gives us a canonical choice function picking one representative from every relational type that is a value of T.

This means that for any concept of pseudoset we have the possibility of erecting an (implemented) theory of relational types of pseudosets. Set theorists are familiar with the extraction of what they call the *arithmetic of T* from a set theory T. This means one can also extract what one can analogously call the *set theory of T*. The set theory of T is of course a theory in the same syntax of T, and it makes sense to ask whether or not the se theory of T is the same as T itself. In unpublished work Holmes has referred to this phenomenon as **reflection**. (This is not to be confused with 'reflection' as in initial segments of the universe "reflecting" the universe in various aspects!). An illustration may help. In ZF we can reason about the relational types of wellfounded extensional relations with a designated element and show that, with the obvious embedding relation they form a model of ZF. Similarly, reasoning in ZF + AFA we can show that the relational types of APGs with the obvious embedding relation give us a model of ZF + AFA.

It is to be hoped that this important topic will receive more attention in the future.

Chapter 6

Ordinals

In this chapter we consider interpretations of set theory plus ordinal arithmetic into set theory. We will start by developing a naïve theory of wellorderings. Wellorderings are not *prima facie* sets, just as ordered pairs are not *prima facie* sets, and we have the same decision to make as we had there, namely between adding to set theory new primitive notations with typed new axioms on the one hand, and fixing one particular implementation on the other. Setting up cardinal arithmetic was a process that didn't involve lots of extra gadgetry, as were able to use Henkin quantifiers to express equipollence. We can use Henkin quantifiers again here, to express isomorphism between wellorderings, but the problem remains of expressing what it is for a structure to be a wellordering in the first place. It seems that the best that can be done is this. We can say that $\mathcal{X} \subseteq \mathcal{P}(X)$ is a wellordering of X iff

1: $(\forall X', X'' \in \mathcal{X})(X' \subseteq X'' \vee X'' \subseteq X')$;
2: $\bigcup \mathcal{X} = X$;
3: $(\forall x_1 x_2 \in X)(x_1 = x_2 \longleftrightarrow (\forall X' \in \mathcal{X})(x_1 \in X' \longleftrightarrow x_2 \in X'))$;
4: $(\forall \mathcal{X}' \subseteq \mathcal{X})(\bigcap \mathcal{X}' \in \mathcal{X})$.

We can write $x_1 <_{\mathcal{X}} x_2$ if $(\exists X' \in \mathcal{X})(x_1 \in X' \wedge x_2 \notin X')$. The idea is that $\mathcal{X}$ is to be the set of all initial segments of the wellordering being coded. Clause (1) says that the order is total; clause (2) says it is defined on the whole of X, clause (3) says it is antisymmetrical. These clauses together say that $\mathcal{X}$ is a strict total order of X. Clause (4) says it is wellfounded. (Every set of initial segments has a smallest element.)

We can then use Henkin quantifiers to say that two wellorderings are isomorphic:

$$\begin{matrix}(\forall x_1 \in X)(\exists y_1 \in Y) \\ (\forall y_2 \in Y)(\exists x_2 \in X)\end{matrix}(x_1 \leq_{\mathcal{X}} x_2 \longleftrightarrow y_1 \leq_{\mathcal{Y}} y_2)$$

6.1 The elementary theory of wellorderings

Recall from p. 12 that capitalised *fraktur* font variables will range over *structures*. In this chapter they will range specifically over *wellorderings*. In keeping with the spirit of the convention there, when it is clear from context that a wellordering is intended, $\mathfrak{X}$ will be the wellordering $\langle X, <_X \rangle$, the structure whose domain is X and whose graph is a relation $<_X$ that wellorders X. (We will not use the calligraphic letter '$\mathcal{X}$' in this context any further: it was used to denote the set of initial segments of a wellorder adn we are not going to allude any further to the fact that that is how wellorderes are to be coded.)

Definition 6.1 If $\mathfrak{X} = \langle X, <_X \rangle$ and $\mathfrak{Y} = \langle Y, <_Y \rangle$ are strict partial orderings then we say $\langle X, <_X \rangle$ **is an end-extension of** $\langle Y, <_Y \rangle$ if $X \subseteq Y$ and $(\forall y \in (Y \setminus X))(\forall x \in X)(\neg(y <_X x))$, and we will write this as $\mathfrak{X} \subseteq_e \mathfrak{Y}$.

We will be interested in this definition primarily in the case where $\langle X, <_X \rangle$ and $\langle Y, <_Y \rangle$ are wellorderings, but there is no harm in giving this more general definition. (In the specific representations of wellorderings given at the start of this chapter the end-extension relation between wellorderings is manifested as set inclusion.)

Definition 6.2 We define the class of all wellorderings as the intersection of all classes of total strict orderings closed under unions of chains (where the order relation is end-extension) and additions of one extra element.

Of course it is more usual (and, mostly, more useful) to say that a relation R is a wellordering if it is a wellfounded strict total order.

By induction on the datatype everything in it is a wellordering. The converse is a bit harder! Suppose $\mathfrak{X} = \langle X, <_X \rangle$ is a wellordering that does not belong to the datatype. Consider the set of $x \in X$ such that the initial segment of $\mathfrak{X}$ bounded by x is a wellordering that is not in the datatype. The closure conditions of the datatype ensure that this set has no minimal element. It is therefore empty. So $\mathfrak{X}$ was minimal such that $\mathfrak{X}$ is a wellordering that does not belong to the datatype. But then it is obtained from initial segments of itself (all of which are in the datatype) by means of operations under which the datatype is closed.

Each of these two definitions can on its own justify a principle of induction over wellorderings. This principle takes two forms, one arising from each definition. Since the datatype of wellorderings is a recursive datatype we deduce an induction principle for it in an obvious way. On the other hand there is a principle of wellfounded induction (transfinite induction, strong induction, course-of-values induction) that we can prove for each individual wellordering.

If $\mathfrak{X} = \langle X, <_X \rangle$ is a wellordering, and P a property such that $(\forall x \in X)(\forall y)(y <_X x \rightarrow P(y)) \rightarrow P(x))$, then $(\forall x \in X)(P(x))$.

This follows immediately by considering the least element (if any) of $\{x \in X : \neg P(x)\}$

Given $\mathfrak{X} = \langle X, <_X \rangle$ and $\mathfrak{Y} = \langle Y, <_Y \rangle$, both wellorderings, we construct the following recursively defined set.

Definition 6.3

(1) $\underline{\hookrightarrow}_{\mathfrak{X}\rightarrow\mathfrak{Y}}$ is to be the $\subseteq$-smallest bijection pairing the $<_X$-first member of X with the $<_Y$-first member of Y) and closed under the following operation: if $X' \subseteq X$ and X' is mapped 1-1 onto $Y' \subseteq Y$ by $\underline{\hookrightarrow}_{\mathfrak{X}\rightarrow\mathfrak{Y}}$ then $\underline{\hookrightarrow}_{\mathfrak{X}\rightarrow\mathfrak{Y}}$ also pairs $x_{X'}$ with $y_{Y'}$ where $x_{X'}$ is the $<_X$-first element of $X \setminus X'$ and $y_{Y'}$ is the $<_Y$-first member of $Y \setminus Y'$.
(2) If $\underline{\hookrightarrow}_{\mathfrak{X}\rightarrow\mathfrak{Y}}$ is defined on the whole of X we write $\mathfrak{X} \underline{\hookrightarrow} \mathfrak{Y}$.
(3) If $\underline{\hookrightarrow}_{\mathfrak{X}\rightarrow\mathfrak{Y}}$ is defined on the whole of X but is not onto Y we write $\mathfrak{X} \hookrightarrow \mathfrak{Y}$.

Theorem 6.1 *Given any two wellorderings, there is a canonical map from one to an initial segment of the other.*

Proof:

It is an immediate consequence of this definition that anything in X that $<_X$-precedes anything in the domain of $\underline{\hookrightarrow}_{\mathfrak{X}\rightarrow\mathfrak{Y}}$ is also in the domain of $\mathfrak{X}$, and Y similarly. The only way in which this construction can fail to eat up all of X and Y is if at some stage the $X' \subseteq X$ we are considering, or the $Y' \subseteq Y$ we are considering, turn out to be empty. If this happens, we have an isomorphism from one to an initial segment of the other. If it never happens, then $\langle X, <_X \rangle$ and $\langle Y, <_Y \rangle$ are isomorphic. ■

Definition 6.4 A structure is **rigid** iff it has no nontrivial automorphisms.

Theorem 6.2 *All wellorderings are rigid.*

Proof:

Suppose $\mathfrak{X}$ is not rigid and let x be the $<_X$-minimal member of X that is moved by an automorphism. So for some automorphism π we have

$x < \pi(x)$, whence $\pi^{-1}(x) < x$. But now $\pi^{-1}(x)$ is smaller than x and is moved by an automorphism, contradicting minimality of x.

■

Corollary 6.1 *Any isomorphism between two wellorderings $\mathfrak{X}$ and $\mathfrak{Y}$ is unique.*

Proof:

If we had two distinct isomorphisms f and g between $\mathfrak{X}$ and $\mathfrak{Y}$ then $f \circ g^{-1}$ would be a nontrivial automorphism of $\mathfrak{Y}$. ■

Definition 6.5 We say $\langle X, <_X \rangle$ **canonically injects into** $\langle Y, <_Y \rangle$ if the canonical bijection $\underline{\hookrightarrow}_{\mathfrak{X}\to\mathfrak{Y}}$ uses up all of $\langle X, <_X \rangle$ and we write $\langle X, <_X \rangle \underline{\hookrightarrow} \langle Y, <_Y \rangle$.

Thus clearly $\langle X, <_X \rangle$ canonically injects into $\langle Y, <_Y \rangle$ iff $\langle X, <_X \rangle \underline{\hookrightarrow} \langle Y, <_Y \rangle$.

Proposition 6.1 $\underline{\hookrightarrow}$ *is transitive.*

Proof: Compose the maps. ■

Proposition 6.1 tells us that $\hookrightarrow$ is a preorder. As before (page 34) we can extract an equivalence relation from this preorder. In fact, we can give two equivalent definitions of the equivalence relation. Clause 2 is what arises by analogy with the discussion on page 34.

Definition 6.6

We write '$\mathfrak{X} \simeq \mathfrak{Y}$' for either of the following:

(1) The canonical bijection $\underline{\hookrightarrow}_{\mathfrak{X}\to\mathfrak{Y}}$ is total and onto;
(2) $\mathfrak{X} \underline{\hookrightarrow} \mathfrak{Y} \wedge \mathfrak{Y} \underline{\hookrightarrow} \mathfrak{X}$.

We say of two wellorderings thus related that they are **of the same length**.

We had better show that these two clauses are equivalent.

$1 \to 2$. The first conjunct is immediate. The second comes from the fact that the inverse of a canonical bijection that is iso is also a canonical bijection.

$2 \to 1$. The composition of two canonical bijections is another canonical bijection. So $\underline{\hookrightarrow}_{\mathfrak{X}\to\mathfrak{Y}} \circ \underline{\hookrightarrow}_{\mathfrak{Y}\to\mathfrak{X}}$ is $\underline{\hookrightarrow}_{\mathfrak{X}\to\mathfrak{X}}$. But this is onto X, so $\underline{\hookrightarrow}_{\mathfrak{X}\to\mathfrak{Y}}$ must have been onto Y. This is 1.

It is true that in this development we have not taken isomorphism as a primitive of this language, because it is convenient to approach it via the uniqueness theorem, corollary 6.1, but let us for the moment imagine we had taken it as primitive. We can then write $\mathfrak{X} \underline{\hookrightarrow} \mathfrak{Y}$ if $(\exists \mathfrak{X}')(\exists \mathfrak{Y}')(\mathfrak{X}' \simeq$

$\mathfrak{X} \wedge \mathfrak{Y}' \simeq \mathfrak{Y} \wedge \mathfrak{X}' \subseteq_e \mathfrak{Y}')$. This is exactly parallel to our definition of $\hookrightarrow_c$ as a relation on *sets* when we were developing cardinal arithmetic: $x \hookrightarrow_c y \longleftrightarrow (\exists x'y')(x \sim x' \wedge y \sim y' \wedge x' \subseteq y')$. This relation is a preorder, and we can extract an equivalence relation from it as usual, and that equivalence relation is—the relation $\simeq$ we first thought of. This is the same state of affairs we found with cardinal arithmetic.

We end up where we started because of the Schröder-Bernstein theorem (as it is known in the cardinal case). The analogous statement for the ordinal version of $\hookrightarrow$ is much more trivial and has just been proved in the discussion following definition 6.6.

Lemma 6.1 *$\simeq$ is an equivalence relation.*

Proof:

(i) $\simeq$ is reflexive because of the identity map.

If $\mathfrak{X} = \langle X, <_X \rangle$ we prove by induction on $<_X$ that $\underline{\hookrightarrow}_{\mathfrak{X} \to \mathfrak{X}}$ is the identity. This is because the identity relation restricted to X is one of the family of bijections of which $\underline{\hookrightarrow}_{\mathfrak{X} \to \mathfrak{X}}$ is defined to be the least.

(ii) $\simeq$ is transitive. (compose the maps)

(iii) $\simeq$ is symmetrical. (take the inverse)

■

The emergence of this isomorphism relation enables us to say what a virtualist analysis of ordinal arithmetic is.

> **(First order) Ordinal arithmetic is the study of those relations between wellorderings for which $\simeq$ is a congruence relation.**

Higher-order ordinal arithmetic will be the study of those relations between wellorderings for which $\langle \simeq_n : n \in \mathbb{N} \rangle$ is a suite of congruence relations, where the $\langle \simeq_n : n \in \mathbb{N} \rangle$ suite will be defined below by analogy with cardinal arithmetic.

Lemma 6.2 *No two distinct initial segments of a wellordering are of the same length.*

Proof: We will prove the following assertion by $<_X$ induction on 'x':

$$(\forall y)(((\langle \{z \in X : z <_X y\}, <_X \rangle \simeq \langle \{z \in X : z <_X x\}, <_X \rangle) \longleftrightarrow x = y)$$

Pick x in X $<_X$-minimal so that there is y in X such that $x \neq y$ but $\langle \{z \in X : z <_X y\}, <_X \rangle \simeq \langle \{z \in X : z <_X x\}, <_X \rangle$. Then pick y minimal so that $x \neq y$ but $\langle \{z \in X : z <_X y\}, <_X \rangle \simeq \langle \{z \in X : z <_X x\}, <_X \rangle$. By hypothesis, x and y are distinct, so one must be $<_X$ the other. Suppose it is x, without loss of generality. But then the initial segment bounded by x is isomorphic to two distinct initial segments of $\mathfrak{X}$ contradicting corollary 6.1.

Lemma 6.3 $\hookrightarrow$ *is wellfounded.*

Proof:

Suppose A is a nonempty set of wellorderings such that no member of it injects into all the others. Let $\mathfrak{X} = \langle X, <_X \rangle$ be an arbitrary member of A. Since A has no element that injects into all others, there are at least some $\mathfrak{Y} = \langle Y, <_Y \rangle$ such that when we construct the canonical injection $\underline{\hookrightarrow}_{\mathfrak{Y}\to\mathfrak{X}}$ from $\mathfrak{Y}$ to $\mathfrak{X}$, there are bits of X that are not in the range of the canonical bijection. Let X' be the collection of elements x of X such that, for some $\mathfrak{Y}$, x is not in the range of the canonical injection $\underline{\hookrightarrow}_{\mathfrak{Y}\to\mathfrak{X}}$.

We will show that X' has no least member under $<_X$. Suppose it does, and x is the $<_X$-least element of X'. Then, for some $\mathfrak{Y} \in A$, x is the first thing not in the range of $\underline{\hookrightarrow}_{\mathfrak{Y}\to\mathfrak{X}}$. But then $\mathfrak{Y}$ injects into every wellordering in A, contradicting the assumption that there is no such $\mathfrak{Y}$.

■

More graphically (because of the connexity of $\hookrightarrow$ (theorem 6.1)) every nonempty set X of wellorderings has a member that canonically injects into all members of X.

Now let us demonstrate a few elementary facts about wellorderings.

Remark 6.1 *If there is an order-preserving embedding $\pi : \mathfrak{X} \to \mathfrak{Y}$ then $\mathfrak{X}$ canonically injects into $\mathfrak{Y}$.*

Proof: We know that $(\forall x \in X)((\underline{\hookrightarrow}_{\mathfrak{X}\to\mathfrak{Y}})(x) \leq \pi(x))$ because $(\underline{\hookrightarrow}_{\mathfrak{X}\to\mathfrak{Y}})(x)$ is the *least* thing in Y not in the range of $\underline{\hookrightarrow}_{\mathfrak{X}\to\mathfrak{Y}}$ restricted to $\{w \in X : w <_X x\}$, whereas all we know about $\pi(x)$ is that it is one of the things in Y not in the range of $\underline{\hookrightarrow}_{\mathfrak{X}\to\mathfrak{Y}}$ restricted to $\{w \in X : w <_X x\}$. ■

Remark 6.1 actually characterises wellorderings in the sense that

Remark 6.2 *A linear ordering $\mathfrak{X}$ is a wellordering iff every linear order that can be embedded in $\mathfrak{X}$ is isomorphic to an initial segment of $\mathfrak{X}$.*

One direction is easy: any subset X' of a wellordered set X' is wellordered by a restriction of the wellordering $<_X$ and $\langle X', <_X \rangle$ will be isomorphic to an initial segment of $\langle X, <_X \rangle$.

Conversely if every subordering is isomorphic to an initial segment we reason as follows. There are subsets that are singletons, so there is a first element. That means that every subset must have a first element, which is to say the ordering is a wellordering.

There are other kinds of injections that we will need to consider later. For example, the function $\lambda n.(2 \cdot n)$ is an injection from $\mathbb{N}$ into itself, but it is not the canonical injection. (That is of course the identity.) It is an example of a **cofinal embedding**.

Definition 6.7 If $f : \mathfrak{X} \to \mathfrak{Y}$ is an order-preserving injection such that $(\forall y \in Y)(\exists x \in X)(y <_{\mathfrak{Y}} f(x))$ then we say that f **cofinally embeds** $\mathfrak{X}$ into $\mathfrak{Y}$, and we write $\mathfrak{X} \hookrightarrow_{cof} \mathfrak{Y}$.

Remark 6.3 $\hookrightarrow_{cof}$ *is transitive.*

Proof: Compose the maps. ■

Definition 6.8

(i) If $\mathfrak{A}$ and $\mathfrak{B}$ are wellorderings of disjoint sets then $\mathfrak{A} \frown \mathfrak{B}$ is the result of concatenating $\mathfrak{B}$ on the end of $\mathfrak{A}$.

(ii) If $\mathfrak{A}$ and $\mathfrak{B}$ are wellorderings then $\mathfrak{A} \times_{colex} \mathfrak{B}$ is the colex product of $\mathfrak{A}$ and $\mathfrak{B}$. ($\langle a, b\rangle < \langle a', b'\rangle$ iff $b < b'$ or $b = b' \wedge a < a'$.)

(iii) If $\mathfrak{A}$ and $\mathfrak{B}$ are wellorderings with 0 the bottom element of $\mathfrak{A}$ then $\mathfrak{A}^{\mathfrak{B}}$ is the wellordering whose domain is the set of all functions from B to A which take the value 0 at all but finitely many arguments, ordered colex (by last difference).

Theorem 6.3 $\simeq$ *is a congruence relation for the relations (i)* $\underline{\hookrightarrow}$, *(ii)* $\hookrightarrow_{cof}$, *and (iii)* $\hookrightarrow$. *It is also a congruence relation for the binary functions (iv)* $\frown$, *(v)* $\times_{colex}$ *and (vi) exponentiation (as in definition 6.8 part (iii)).*

Proof:

(i) We need to show that if $\mathfrak{X} \simeq \mathfrak{X}'$, $\mathfrak{Y} \simeq \mathfrak{Y}'$ and $\mathfrak{X} \underline{\hookrightarrow} \mathfrak{Y}$ then $\mathfrak{X}' \underline{\hookrightarrow} \mathfrak{Y}'$. Now $\mathfrak{X} \simeq \mathfrak{X}'$ tells us there is a map $\pi : \mathfrak{X} \simeq \mathfrak{X}'$; $\mathfrak{X} \underline{\hookrightarrow} \mathfrak{Y}$ tells us there is a map $\tau : \mathfrak{X} \underline{\hookrightarrow} \mathfrak{Y}$ and $\mathfrak{Y} \simeq \mathfrak{Y}'$ tells us there is a map $\sigma : \mathfrak{Y} \simeq \mathfrak{Y}'$. The composition $\sigma \circ \tau \circ \pi$ is a map $\mathfrak{X}' \underline{\hookrightarrow} \mathfrak{Y}'$. The proofs of (ii) and (iii) are analogous. The remaining parts are an unilluminating grind left to the reader. ■

Finally we define $\simeq_n$, for $n = 1, 2, \ldots$ on sets of wellorderings, sets of sets of wellorderings ... as in section 3.3.

6.2 The language of ordinal arithmetic

As before, we will write '$\mathcal{L}^*$' for the new, extended language, and $\mathcal{I}$ for the interpretation. This time it is set theory with ordinal arithmetic. It will have the following items of syntax. (i) binary relations $\leq_{On}$ and $<_{On}$ arising from $\underline{\hookrightarrow}$; (ii) a function letter written cf, for 'cofinality'. (This is because $\simeq$ is a congruence relation for $\hookrightarrow_{cof}$ (this is theorem 6.3) and the fact that for any wellordering there is—up to a canonical isomorphism—a unique shortest wellordering $\hookrightarrow_{cof}$); (iii) two binary functions written with infix notation: $\alpha = \beta + \gamma$, $\alpha = \beta \times \gamma$, and an infinite family of predicate letters $=_{On}$, $=_1$, $=_2 \ldots$, We can also add a notation $\ulcorner \alpha = otp(\mathfrak{X}) \urcorner$, which says that α is the order type of $\mathfrak{X}$.

We now define $\mathcal{I} : \mathcal{L}^* \to \mathcal{L}$.

Definition 6.9

(1) $\mathcal{I}$ of a Greek variable is the corresponding upper-case fraktur variable with the subscript doubled;
(2) $\mathcal{I}$ of an upper-case fraktur variable with subscript i is the upper-case fraktur variable with the same body and subscript $2i + 1$;
(3) $\mathcal{I}$ of '$=_{On}$' is '$\simeq$';
(4) $\mathcal{I}$ of '$=_n$' is '$\simeq_n$' for $n \geq 1$;
(5) $\mathcal{I}$ of '$\leq_{On}$' is '$\underline{\hookrightarrow}$';
(6) $\mathcal{I}$ of '$<_{On}$' is '$\hookrightarrow$';
(7) $\mathcal{I}$ of $\ulcorner otp(\mathfrak{X}_i) = \alpha_j \urcorner$ is $\ulcorner \mathfrak{X}_{2i+1} \simeq \mathfrak{A}_{2j} \urcorner$;
(8) $\mathcal{I}$ of $\ulcorner \gamma = \alpha + \beta \urcorner$ is $\ulcorner (\exists \mathfrak{A}' \mathfrak{B}')(\mathfrak{A} \simeq \mathfrak{A}' \wedge \mathfrak{B} \simeq \mathfrak{B}' \wedge (A \cap B = \emptyset) \wedge \mathfrak{A}' \frown \mathfrak{B}' \simeq \mathfrak{C}) \urcorner$;
(9) $\mathcal{I}$ of $\ulcorner \alpha = \beta \times \gamma \urcorner$ is $\ulcorner \mathfrak{A} \simeq \mathfrak{B} \times_{colex} \mathfrak{C} \urcorner$;
(10) $\mathcal{I}$ of $\ulcorner \alpha = \beta^{\gamma} \urcorner$ is $\ulcorner \mathfrak{A} \simeq \mathfrak{B}^{\mathfrak{C}} \urcorner$.

This list is not exhaustive: remember that $\mathcal{L}^*$ contains a symbol for every relation or function (atomic or molecular) of $\mathcal{L}$ for which $\{\simeq_n : n \geq 1\}$ is a suite of congruence relations.

We will pass over first-order ordinal arithmetic fairly swiftly. It tells us that addition and multiplication and exponentiation are not commutative, that they distribute on one side but not the other, and so forth. This is standard material and does not need rehearsal here.

Even higher-order ordinal arithmetic has little to detain us, although there is a fundamental result which we will have to prove, namely that $<_{On}$ is a wellordering of ordinals. Experience with other kinds of order types would not lead us to expect this. The embedding relation between order types of total orders is not a total order (think of ω and ω^*) or even antisymmetrical (think of the rationals with and without endpoints).

Theorem 6.4 *$<_{On}$ is a wellordering*.*

Proof: To prove that $<_{On}$ is a wellordering* we need to know that it is transitive, trichotomous and wellfounded.

$\mathcal{I}$ of the assertion that $<_{On}$ is transitive is

$$(\forall \mathfrak{X})(\forall \mathfrak{Y})(\forall \mathfrak{Z})((\mathfrak{X} \hookrightarrow \mathfrak{Y} \wedge \mathfrak{Y} \hookrightarrow \mathfrak{Z}) \to \mathfrak{X} \hookrightarrow \mathfrak{Z})$$

which follows immediately from the transitivity of $\hookrightarrow$, remark 6.1.

$\mathcal{I}$ of the assertion that $<_{On}$ is trichotomous is

$$(\forall \mathfrak{X} \forall \mathfrak{Y})((\mathfrak{X} \hookrightarrow \mathfrak{Y}) \vee (\mathfrak{Y} \hookrightarrow \mathfrak{X}) \vee \mathfrak{X} \simeq \mathfrak{Y})$$

which (more or less) is Theorem 6.1.
$\mathcal{I}$ of the assertion that $<_{On}$ is wellfounded is:

$(\forall \text{ sets } \Xi \text{ of ordinals })(\Xi \neq_1 \emptyset \rightarrow (\exists \alpha \in_1 \Xi)(\forall \beta \in_1 \Xi)(\beta \not<_{On} \alpha))$

but since we have just shown that $<_{On}$ is total this is

$(\forall \text{ sets } \Xi \text{ of ordinals })(\Xi \neq_1 \emptyset \rightarrow (\exists \alpha \in_1 \Xi)(\forall \beta \in_1 \Xi)(\alpha \leq_{On} \beta))$

eliminating Greek variables by applying $\mathcal{I}$ (and then relettering)

$(\forall \text{ sets } X \text{ of wellorderings })(X \neq \emptyset \rightarrow (\exists \mathfrak{X} \in X)(\exists \mathfrak{X}' \simeq \mathfrak{X})(\forall \mathfrak{Y} \in X)(\exists \mathfrak{Y}' \simeq \mathfrak{Y})(\mathfrak{X}' \underline{\hookrightarrow} \mathfrak{Y}'))$

which is equivalent to
$(\forall \text{ sets } X \text{ of wellorderings})(X \neq \emptyset \rightarrow (\exists \mathfrak{X} \in X)(\forall \mathfrak{Y} \in X)(\mathfrak{X} \underline{\hookrightarrow} \mathfrak{Y}))$
and this is of course lemma 6.3.

■

6.3 Ordinals of wellorderings of sets of ordinals

Very well, we have seen that each set of ordinals is naturally wellordered by $<_{On}$, inside scare-quotes at least. Let's think a bit about ordinals of wellorderings-of-sets-of-ordinals-under-$<_{On}$. These ordinals will be doubly virtual, and before we can get our hands on them we must ensure that we have—over the (singly virtual) ordinals—all the gadgetry that we needed in $\mathcal{L}$ to get virtual ordinals in the first place. That was pairing. So we need to know what a wellordering of a set of ordinals is, and before that we need to know what a relation on a set of ordinals is, and before *that* we need to take at least a sideways glance at what an ordered pair of ordinals might be, even if we decide to run up the white flag and use branching quantifiers instead.

We can use Hinnion's '+' operator to give us equality on sets-of-widgets once we have equality for widgets, and the equivalence relation that is appropriate for ordered pairs of widgets is: "$\langle x, y \rangle$ is equivalent to $\langle x', y' \rangle$ as long as x is equivalent to x' and y is equivalent to y'". (We saw an earlier instance of this on page 54.) We can use these two recursively to define equality for sets of ordinals, ordered pairs of ordinals, sets of ordered pairs of ordinals and so on. For example, $\mathcal{I}$ of a variable ranging over wellorderings of sets of ordinals is a variable ranging over sets of ordered pairs $\langle x, y\rangle$, $\langle x', y'\rangle$ ... such that all the xs are identical as sets of ordinals and all the ys are identical as sets of pairs of ordinals.

In higher-order ordinal arithmetic we are free to consider sets of ordinals ordered according to all sorts of principles, but in fact the only way in which we will be wellordering sets of ordinals here will be according to $<_{On}$.

Ordinals of wellorderings-of-sets-of-ordinals-under-$<_{On}$ are doubly virtual: one takes one step to be realist about ordinals over wellorderings of sets, and even then ordinals of wellorderings-of-sets-of-ordinals-under-$<_{On}$ remain virtual. This means that the best way to describe what is going on is in terms of a *third* language $\mathcal{L}^{**}$ and an interpretation $\mathcal{I}^*$ from $\mathcal{L}^{**}$ into $\mathcal{L}^*$. Notice that what we are doing is *not* the same as supposing that $\mathcal{L}^*$ already contained ordinal arithmetic of higher order. We made that assumption to start with! (See definition 6.9.) Ordinals of wellorderings of sets of ordinals are not remotely the same as sets of sets of ordinals. The first are doubly virtual: the second are singly virtual.

However **wellorderings** of sets of ordinals are not doubly virtual but merely singly virtual. We could have incorporated talk of such entities into $\mathcal{L}^*$ to start with but as it happens we didn't. There is still time to put this right.

Add to $\mathcal{L}^*$ the following features:

(1) a supply of lower case fraktur type variables (which will range over wellorderings of sets of wellorderings);
(2) a predicate: $\mathfrak{a}$ is a wellordering* of Ξ;
(3) a derived one-place predicate "is a wellordering*" and a notion of isomorphism between two wellordering*s of sets of ordinals. We write this last with $\simeq^*$;
(4) Binary relations $\underline{\hookrightarrow}^*$ and $\hookrightarrow^*$ and $\simeq^*$. These are flanked by lower case fraktur type variables. Only the first is primitive: the others are defined in terms of it.

In $\mathcal{L}$ we used upper-case fraktur font variables to range over structures, and we had a convention that if $\mathfrak{X}$ was a wellordering $\mathfrak{X}$ was to be decomposed as $\langle X, <_X\rangle$. (Using the terminology from p. 12, $DM(\mathfrak{X}) = X$ and $DG(\mathfrak{X}) = <_X$.) A similar convention will be in operation here, applicable to lower case fraktur type variables which will range over wellorderings of sets of ordinals, to the effect that $\mathfrak{x}$ is the structure with domain Ξ (a set of ordinals) and a relation $<_\Xi$.

We define $\mathcal{I}$ for these new bits of machinery as follows.

(1) $\mathcal{I}$ of $\mathfrak{a}$ is a wellordering* of Ξ is

"A is a transitive wellfounded relation on X and X is a set of wellorderings $\wedge (\forall \mathfrak{X}, \mathfrak{Y} \in X)(A(\mathfrak{X}, \mathfrak{Y}) \vee A(\mathfrak{Y}, \mathfrak{X}) \vee \mathfrak{X} \simeq \mathfrak{Y})$"[1];

(2) $\mathcal{I}$ of $(\forall \mathfrak{a})(\ldots)$ is: $(\forall A)(\forall R_A)([R_A$ is a wellfounded relation on A and A is a set of wellorderings and $(\forall \mathfrak{X}, \mathfrak{Y} \in A)((\mathfrak{X} R_A \mathfrak{Y}) \vee \mathfrak{Y} R_A \mathfrak{X} \vee \mathfrak{X} \simeq \mathfrak{Y})] \to \mathfrak{I}$ of $\ldots)$;
Existential quantifier similarly

(3) $\mathcal{I}$ of $\mathfrak{a} \underline{\hookrightarrow}^* \mathfrak{b}$ is: there is a relation S whose domain is the whole of A and whose codomain is a subset of B and

  (a) $(\forall \mathfrak{B} \in B)(S^{-1}$"$\{\mathfrak{B}\}$ is a $\simeq$-equivalence class of members of $A)$;
  (b) $(\forall \mathfrak{A} \in A)(S$"$\{\mathfrak{A}\}$ is a $\simeq$-equivalence class of members of $B)$;
  (c) $(\forall \mathfrak{A}, \mathfrak{A}' \in A)(\forall \mathfrak{B}, \mathfrak{B}' \in B)(\mathfrak{A} R_A \mathfrak{A}' \wedge \mathfrak{A} S \mathfrak{B} \wedge \mathfrak{A}' S \mathfrak{B}' \to \mathfrak{B} R_B \mathfrak{B}')$ (two things in A are the same length iff they are related by S to the same things in B);
  (d) $(\forall \mathfrak{B}, \mathfrak{B}' \in B)(\forall \mathfrak{A}, \mathfrak{A}' \in A)(\mathfrak{B} R_B \mathfrak{B}' \wedge \mathfrak{B} S^{-1} \mathfrak{A} \wedge \mathfrak{B}' S^{-1} \mathfrak{A}' \to \mathfrak{A} R_A \mathfrak{A}')$. (two things in B in the codomain of S are the same length iff they are related by S^{-1} to the same things in A.)

$\mathcal{L}^{**}$ will be an extension of $\mathcal{L}^*$ with the following new machinery (roughly any item from the lexicon of $\mathcal{L}^* \setminus \mathcal{L}$ with an asterisk appended).

(1) Upper-case $\mathcal{CALLIGRAPHIC}$ type letters will range over ordinals of wellorderings of sets of ordinals.
(2) $\leq_{On^*}$, $<_{On^*}$, $=_{On^*}$
(3) $=_{n^*}$ for each $n \geq 1$.
(4) There is a unary function $otp^*()$ where the argument is a a lower case fraktur variable and the value is an upper-case calligraphic variable;
(5) We have had our nose rubbed in the fact that expressions like "The set of ordinals below α is wellordered by $<_{On}$ to length α" do not make sense. However the type-violating intuition that is rebuked by this discovery deserves to be acknowledged, and we do this by introducing a binary relation written $T(\alpha) = \mathcal{C}$ which arises from $\langle \{\beta : \beta <_{On} \alpha\}, <_{On} \rangle \simeq^* \mathfrak{c}$.

Definition 6.10
We define $\mathcal{I}^* : \mathcal{L}^{**} \to \mathcal{L}^*$ as follows.

(1) $\mathcal{I}^*$ of an upper-case calligraphic variable is the corresponding lower case fraktur variable;
(2) $\mathcal{I}^*$ of $=_{On^*}$ is $\simeq^*$;
(3) $\mathcal{I}^*$ of $=_{n^*}$ is $\simeq_{n^*}$ for $n \geq 1$;

[1]Remember that the objects that correspond to wellorderings of ordinals are not wellorderings of sets of wellorderings, but rather relations on sets of wellorderings which satisfy a weak version of trichotomy: given two wellorderings $\mathfrak{X}$ and $\mathfrak{Y}$ we must have $R(\mathfrak{X}, \mathfrak{Y}) \vee R(\mathfrak{Y}, \mathfrak{X}) \vee \mathfrak{X} \simeq \mathfrak{Y}$ rather than $R(\mathfrak{X}, \mathfrak{Y}) \vee R(\mathfrak{Y}, \mathfrak{X}) \vee \mathfrak{X} = \mathfrak{Y}$.

(4) $\mathcal{I}^*$ of $\leq_{On^*}$ is $\underline{\hookrightarrow}^*$;
(5) $\mathcal{I}^*$ of $<_{On^*}$ is $\hookrightarrow^*$;
(6) $\mathcal{I}^*$ of $\ulcorner otp^*(\mathcal{X}) = \alpha \urcorner$ is $\ulcorner \mathcal{X} \simeq^* \mathcal{A} \urcorner$;
(7) $\mathcal{I}^*$ of $\ulcorner T(\alpha) = \mathcal{C} \urcorner$ is $\ulcorner \langle \{\beta : \beta <_{On} \alpha\}, <_{On} \rangle \simeq^* \mathfrak{c} \urcorner$.

A proposition central to the derivation of Hartogs' theorem and the Burali-Forti paradox is the assertion that every ordinal counts the set of its predecessors in their natural ordering. With the advent of doubly virtual ordinals we can see this for what it is: a pun. However there is an assertion with which it could be confused, and which can in fact be proved in most of the weak set theories that we have not so far had to choose between for our set-theoretic background.

Remark 6.4 *Every wellordering $\mathfrak{X}$ is isomorphic to the set of initial segments of $\mathfrak{X}$ wellordered by end-extension.*

Proof: The assertion that $\mathfrak{X}$ is isomorphic to the set of initial segments of $\mathfrak{X}$ wellordered by end-extension is (or can be written out as) an expression of $\mathcal{L}$ with one free variable, and is thus a (molecular) monadic predicate of $\mathcal{L}$. It is relatively simple to check that this is a predicate for which $\simeq$ is a congruence relation. Thus we can think of it as a property of ordinals. We prove by induction on the wellfounded structure of wellorderings quasi-ordered by end-extension that all wellorderings have this property. We show that if it is true for every initial segment $\langle X', \hookrightarrow_{X'} \rangle$ of $\langle X, \hookrightarrow_X \rangle$ that $\langle X', \hookrightarrow_{X'} \rangle$ is isomorphic to the set of its initial segments ordered by end-extension, then it is true also for $\langle X, \hookrightarrow_X \rangle$. The isomorphism is easy to construct if $\langle X, \hookrightarrow_X \rangle$ has a top element (just add an ordered pair, as it were). If $\langle X, \hookrightarrow_X \rangle$ does not have a top element the desired bijection is simply the union of the bijections, which, by inductive hypothesis, already exist.

■

This assertion cannot be proved in set theories in which there are lingering type distinctions (with respect to which a set is of a different 'type' from its members) such as the Quine systems. I mention this is order to separate the question of its provability in any given system from the provability of the pun, since it is easy for the two to become confused.

6.3.1 *What does the T function do?*

Understanding the behaviour and purpose of the T function and its analogues elsewhere is of central importance in understanding iterated virtuality. Readers should start by putting themselves into a frame of mind in which they think of singly virtual ordinals as real, that is to say real enough for there to be real sets of them so that they (the reader) can thereafter think of the doubly virtual ordinals of wellorderings of sets of ordinals as singly virtual. (In such a frame of mind one is less likely to recklessly identify the two sorts of entity.)

From this semirealist perspective, it is clear that T is an isomorphism between the first-generation (sometime singly virtual but now real) ordinals and the second generation (sometime doubly virtual now singly virtual) ordinals. We can ask whether or not it is actually the identity only once we have settled on implementations of these two kinds of ordinals. We will consider this later. For the moment I want to review some topics in cardinal arithmetic that we can review only now that we have the T-function.

The T-function is defined on cardinals as well, though the definition is not so well motivated. If x is a set of sets all of different sizes, and $|x| = \alpha$, then $T\alpha$ is the size of $\{|y| : y \in x\}$. With this is mind we can now see more clearly the relation between the two ways of defining multiplication, and the two ways of defining exponentiation.

Cartesian product of sets a and b gives rise to an operation on cardinals α and β. Let's write it $\times_1$ for the moment.

The union of a disjoint family A of things each equipollent to b gives rise to another operation, taking $T\alpha$ and β. Let's write it $\beta \times_2 T\alpha$. (I'm writing '$T\alpha$' to remind us that the second argument to $times_2$ must be a value of T—a doubly virtual cardinal.)

There are various algebraic laws relating multiplication to addition and exponentiation, and it is easy to show that $\times_1$ obeys all of them. It is not at all easy to show the same for $\times_2$: it's not even obvious that $\alpha \times_2 T\beta = \beta \times_2 T\alpha$. (This is not a pun: it is well-typed.) Indeed, as we noted earlier, we needed the axiom of choice even to establish that $\times_2$ was well-defined in the first place, let alone to ascertain its value. All the more unfortunate then, that $\times_2$ is the concept of multiplication being used in Lagrange's theorem (see page 47). What Lagrange proved was that if G' is a subgroup of G then $(\exists\beta)(|G'| \times_2 \beta = |G|)$ which (with the axiom of choice, provable for finite sets) is equivalent to $(\exists\beta)(|G'| \times_1 T^{-1}(\beta) = |G|)$.

Similarly we have two definitions of exponentiation.

The set of all functions from a to b gives rise to an operation on cardinals which we have been writing β^α but which we will write $\beta\ exp_1\ \alpha$ for the nonce.

The direct product of a set a of elements all equipollent to b gives rise to an operation on cardinals which we will write $\beta\ exp_2\ \alpha$.

exp_1 is an easy operation to use, because both its arguments are singly virtual, and it is easy to show that it obeys the obvious algebraic laws that one expects. How different is exp_2, which has a singly virtual argument and a doubly virtual argument, and for which we need the axiom of choice to prove that it is even well-defined. Sadly, exp_2 is the concept of exponentiation employed in Euler 's theorem. What Euler proved was that $a\ exp_2 \phi(n) = 1$ (mod n) which (with the axiom of choice, provable for finite sets) is equivalent to $a\ exp_1(T^{-1}\phi(n)) = 1$ (mod n). (p. 57) The functions $\times_2$ and exp_2 have arguments of different types, and one can use T to raise the type of an argument to turn a pun (like $\alpha \times_2 \beta = \beta \times_2 \alpha$) into something legitimate (as above). To return to an earlier example, the T function also enables us to express "x is relatively large" in a pun-free way. That is simply '$|x| \geq T(min(x))$'. (Brief reality check: x is a set of cardinals, so $|x|$ is doubly virtual and the members of x—in particular $min(x)$—are singly virtual, so $T(min(x))$ is doubly virtual, and can be likened or unlikened with $|x|$.)

It is very tempting to argue as follows. T is an isomorphism between singly virtual cardinals and double virtual cardinals, and any isomorphism between mathematical structures just *is* identity. One reply to this could be that it is question-begging: on a virtualist account the singly virtual cardinals don't exist, and nor *a fortiori* do the doubly virtual cardinals, so they aren't around to be isomorphic to anything and the argument about the nature of mathematical objects never gets off the ground. A more significant rebuttal might be that isomorphism is actually not a criterion for identity of mathematical objects anyway, as witness the fact that the finite cardinals and the finite ordinals are isomorphic but are widely regarded as distinct. It is true that the Von Neumann implementation of cardinals and ordinals identifies them, but this is regarded—if not as a bug (a *pretty* bug perhaps)—then certainly not as a feature. And then there are implementations of cardinals and ordinals that implement finite cardinals and finite ordinals differently. Are they defective implementations on that account?

Finally a diehard virtualist might even turn this objection to his advantage. Anyone who says that an idea has not gained the support it merits strengthens their case if they can explain why that support was not forthcoming. One does not need to believe that the distinction between singly and multiply virtual cardinals is spurious to have an explanation for why it does not leap to the eye: the isomorphism T is so obvious that one is tempted by the idea that that it should be the identity. And this temptation—as the example of finite ordinals and finite cardinals shows—is chimærical.

There is always a trade-off between strength and structure: the stronger the theory, the fewer the models. By choosing a stronger theory to examine a phenomenon, one obliterates structure. Stronger theories prove more theorems, and more equations, and make more and more things look equivalent. In passing to a stronger theory we always need to be sure as far as possible that the only structure we thereby obliterate is spurious structure, and that all the genuine structure remains. I used to think that a move from constructive logic to classical logic was secure in that sense, namely that all the differences between classically-equivalent but constructively distinct propositions were spurious. (In those days I was a young philosophy student who had overdosed on Quine.) A situation that was in some ways parallel to the one we face here is Russell and Whitehead's axiom of reducibility which expunged all the structure of ramifications introduced in their type theory. For most purposes it is safe to regard the structure expunged thereby as spurious. What about the structure expunged by equating T to the identity—is that spurious?

Manifestly, the community of set theorists is overwhelmingly of the view that the T function of ordinal arithmetic is the identity. This state of affairs is much more the consequence of the triumph of the Von Neumann implementation in the struggle between implementations for public acceptance (which in turn is a consequence of the undeniable neatness of the Von Neumann implementation) than the outcome of philosophical cogitation about the nature of ordinals. That is not to say that the received view is *incorrect*, merely that it was reached by a process other than a debate about the issues concerned.

It may be worth noting that for many implementations the assertion that T is the identity has nontrivial consistency strength. Without going into too many details, one can at least note that if T is to be the identity then certain sets which might not, *prima facie* be the same size, nevertheless *are* the same size. At the very least this is an assertion about the existence of maps, and therefore might imply nontrivial comprehension axioms. In ZF, and indeed in Zermelo and even the system Mac of Mathias [2002] the set existence axiom needed to equate T and the identity are all theorems anyway. However it is known that if one adds to Quine's NF merely enough comprehension to ensure that the Russell-Whitehead implementation *of the natural numbers* makes T into the identity then the consistency of NF can be proved in the resulting system. But this is recondite! The moral to carry away from these reflections is that if one wishes to have a virtualist account of cardinals (or ordinals) and yet believe the same invariant ("pun-free") sentences of cardinal or ordinal arithmetic as a platonist then one is committed to believing certain set existence axioms. These set existence axioms are in fact all unstratified, and this consideration is the only serious argument against systems like KF and NF which have only stratified axioms.

6.3.2 *Hartogs' theorem*

Hartogs' theorem states that for every set X there is a wellordered set Y that is not equipollent to any subset of X. It is possible to give extremely snappy proofs of this result (one such is in Johnstone (1987) which—even in its entirety—is hardly any longer than the sketch in the next paragraph) but they do not make the ideas clear.

The idea behind the proof is to consider the theory T_X of wellorderings of parts of X. Next consider the theory of ordinals of those wellorderings. Does this theory have an implementation whose graph is a set of the model? If so, consider the set of implemented ordinals of T_X. We know that it is a pun to assert that the set of ordinals below α is naturally wellordered to length α: it's naturally wellordered to length $T\alpha$. However, if the graph of the implementation (the function that sends each wellordering to its ordinal) is a set of the model we can prove by induction on the set of ordinals that T is the identity. This is possible because we can always do induction on wellorderings (see p. 68). Then the set of (implemented) ordinals of wellorderings of subsets of X is, indeed, a wellordered set too long to be a wellordering of any subset of X, since its length is the least ordinal not in it, namely the least ordinal not the length of a wellordering of a subset of X.

What is very disconcerting about this theorem is that although it appears to be, and in some uncontroversial sense clearly *is* a theorem purely about sets (with pairing), *the manner in which it is proved in any axiomatic set theory will depend on how one implements ordinal arithmetic in that theory*. For each set X, there will be one proof of Hartogs' theorem for X for each implementation of T_X whose graph is a set. The sets that feature in these proofs appear there solely in virtue of their rôles in the implementation of ordinal arithmetic that is being used. One has the *trompe l'œil* impression of a counterexample to the interpolation lemma. For example, the proof of Johnstone's alluded to above procedes by directly constructing a von Neumann ordinal of the appropriate cardinality, using the Mostowski collapse. This will not work in Zermelo set theory, for example. In that context one has to consider the set of wellorderings of subsets of X, (this is a set by power set and separation) and then take its quotient under isomorphism (this too is a set by separation and power set). This quotient will serve as a nonce implementation of the ordinals of wellorderings of subsets of X, and we can prove $(\forall\alpha)(\alpha = T\alpha)$ on it. The underlying idea is the same, but the proof looks quite different because of the different implementation of ordinal being used.

Each implementation of ordinals in set theory gives rise to a proof of Hartogs' theorem (subject to the availability of the requisite comprehension axioms), and every proof of Hartogs' theorem arises from an implementation in this way. The reasoning that is efficacious in proving Hartogs' theorem is efficacious only in virtue of being the translation (*via* an implementation) of resoning about ordinals.

6.3.2.1 *Diagonal intersections*

Finally we should acknowledge a construct which is very useful in large cardinal theory but which is not typed in the way a virtualist would like. If $\langle A_\alpha : \alpha < \beta \rangle$ is a sequence of sets of ordinals, then the **diagonal intersection** is $\{\zeta : (\forall \gamma < \zeta)(\zeta \in A_\gamma)\}$. It seems that this idea may have appeared first in Keisler-Tarski [1960].

6.4 Implementations of ordinal arithmetic

Set theory has paradoxes but no puns; virtual ordinal arithmetic has puns but no paradoxes. Implementations, unlike simulations, allow one to give truth-values to puns, namely the truth-values of the expressions in the base language to which implementation maps the pun. This means that any implementation of ordinal arithmetic in set theory is at risk of sending a pun to one of the paradoxes with which set theory is so richly supplied. This is not the same as saying that the puns themselves are paradoxical. The Burali-Forti paradox results from implementations that give truth-values to the pun '$otp(\{\alpha : \alpha < \beta\}) = \beta$', or '$T\beta = \beta$' for short.

If the collection of implemented ordinals is to be a set, then there must be a failure of induction, for we can prove the pun $otp(\{\alpha : \alpha < \beta\}) = \beta$ by induction on β. This should not be looked upon as a pathology: failures of induction are both deep and natural, as the incompleteness of any recursively axiomatisable arithmetic reminds us. Also, as we shall see below (page 88) induction is inherently paradoxical.

There is one general point that can be made about implementations of ordinal arithmetic before we move on to specific implementations. Suppose we have a faithful implementation whose graph is a set. Then the collection of all implemented ordinals is a set. Can we just concatenate these wellorderings to deduce an inconsistency? The idea is that there can be no longest wellordering because one can always obtain from any wellordering a longer one by removing the first element and putting it on the end. Well, we can if the implementation of β is an end-extension of the implementation of α whenever $\beta > \alpha$ (or strictly we take the nested union of them). But this order-preserving condition is very strong indeed, and only very nice implementations like Von Neumann ordinals have anything like it (see footnote p. 33.) The existence of implementations of cardinal arithmetic that satisfy a condition like this implies weak versions of the axiom of choice (see p. 34.)

If Hartogs' theorem holds, so there is no largest cardinal, then one can show by elementary means that every wellordering of length α has an end-extension of length β for any $\beta > \alpha$. If we have DC_Ω (the axiom of dependent choices for all transfinite sequences) then we would be able to pick an order preserving set of representatives and obtain a longest wellordering.

Thus unrestricted dependent choices is incompatible with a faithful implementation as long as it is a set. (Unless there is a largest wellordered set.) There is no reason why one can't have a faithful implementation: there is one in NFU + AC for example, where one can pick a representative from each isomorphism class of wellorderings. However this implementation—although setlike—is not a set, and in NFU+AC there is a largest wellorderable set.

Let's have a look at some of these implementations.

6.4.1 *The Russell-Whitehead implementation and Scott's trick*

The Russell-Whitehead implementation is the obvious one: take the ordinal of a wellordering to be the isomorphism class of that wellordering. In set theories with a sumset axiom this results fairly quickly in the universe being a set (although the details depend on how wellorderings are represented and possibly on how ordered pairs are implemented), and this is incompatible with most schemes of separation that one might dream up. However it is perfectly feasible in typed set theories like that of *Principia Mathematica.*

The only untyped set theory in which the Russell-Whitehead implementation of ordinals is known to be completely unproblematic is the Jensen-Quine system NFU. In NFU the collection of all (Russell-Whitehead) ordinals is a set and is naturally wellordered in the obvious way.

In NFU (and in NF, for that matter), the comprehension axioms are restricted to formulæ that are stratified. This has the effect that in general it cannot be proved that x and $\{\{y\} : y \in x\}$ are the same size, and the map that sends $|x|$ to $|\{\{y\} : y \in x\}|$ has always been denoted 'T'. T is *inhomogenous*—its argument and value are of different (set-theoretic) types. It is often said that it is the weakness of comprehension that gives rise to this T function. This is a truth, but a confusing truth. The root of the trouble is the fact that the Russell-Whitehead implementation of cardinal is inhomogeneous, in the set-theoretic sense that "$\alpha = |x|$" is stratified with 'α' one level higher than 'x'. What is going on is that the T function in cardinal arithmetic, which is inhomogeneous in the *different* sense that it takes singly virtual cardinals to doubly virtual cardinals is implemented by a set theoretic function which is inhomogeneous in the NF sense, namely its argument and its value are of different *set-theoretic* types. Thus a singly virtual/doubly virtual distinction gets implemented in NF as a first-order/higher-order distinction.

In Quine's ML (original version) the Russell-Whitehead implementation is neither faithful nor is a set, but T is implemented as the identity, and the collection of all Russell-Whitehead ordinals is a set. This is enough to imply a contradiction. T comes to be implemented as the identity because the notion of wellordering in this version of ML is that where every sub**class** of the domain has a bottom element. Wellorderings satisfying this strong condition are sufficiently well behaved for one to be able to prove by induction on their ordinals that T is the identity. In addition, the comprehension axioms are lax enough for the collection of isomorphism classes of these objects to be a set. In the revised version the class of strict total order all of whose subclasses have a minimal element is not longer a set. The collection of strict total orders all of whose subsets have least elements is a set, but wellorderings in that sense are not sufficiently well-behaved for us to be able to prove by induction on their ordinals that T is the identity.

6.4.1.1 *Scott's trick*

We noted above that the Russell-Whitehead implementation is blocked in set thoeries without a universal set. However there is a version of the Russell-Whitehead implementation which can be run by means of a device sometimes known as *Scott's trick*. Take the ordinal of a wellordering to be the set of wellorderings of minimal rank that are isomorphic to that wellordering. Of course Scott's trick works in any set theory with foundation. It even works in any set theory (like ZF + the antifoundation axiom of Forti and Honsell) in which every set is merely the same size as a well-founded set. This is Coret's [1964] axiom B. Of course it is applicable to areas other than cardinal arithmetic: it is used to give an internal treatment of ultrapowers of the universe in large cardinal theory for example. Thus Coret's axiom B enables us to show that any theory of virtual objects arising from a congruence relation has an implementation. Given the nice metamathematical properties of this axiom (see for example Forster [2003]) it might be worth trying to state and prove a converse.

However the implementations given by Scott's trick are not faithful.

6.4.2 *The Von Neumann implementation*

First we define VN, which is to be the collection of all (implemented) ordinals.

(1) The empty set is in VN;
(2) If x is in VN so is $x \cup \{x\}$;
(3) VN is closed under unions of chains.

More formally, VN is the intersection of all things which contain the empty set, are closed under $\lambda x.(x \cup \{x\})$ and unions of chains. It is routine that all sets generated in this way will be wellfounded. (It's a least fixed point and least fixed points are wellfounded.)

Then we have to invoke the Mostowski collapse lemma. The Mostowski collapse lemma asserts *inter alia* that for every wellfounded binary structure $\langle X, R\rangle$ there is a unique transitive set y such that $\langle y, \in\rangle \simeq \langle X, R\rangle$. Let us write "$\langle y, \in\rangle = mtk(\langle X, R\rangle)$" to describe this state of affairs. It is well-known that the Mostowski collapse lemma makes unavoidable use of the axiom scheme of replacement: there are subsystems of ZF which are too weak to prove it and are nevertheless of interest. See Mathias [2001] for a survey. A special case of the Mostowski collapse lemma states that (i) every wellordering is isomorphic to a von Neumann ordinal, (ii) $\mathfrak{X} \simeq mtk(\mathfrak{X})$ and that (iii) $mtk(\mathfrak{X}) \in VN$.

Then we define $\mathcal{I}$.

$\mathcal{I}$ of $\ulcorner(\forall\alpha)(....)\urcorner$ is $\ulcorner(\forall a \in VN)(\mathcal{I}$ of $....)\urcorner$;
$\mathcal{I}$ of $\ulcorner(\exists\alpha)(....)\urcorner$ is $\ulcorner(\exists a \in VN)(\mathcal{I}$ of $....)\urcorner$;
$\mathcal{I}$ of $\ulcorner\alpha < \beta\urcorner$ is $\ulcorner a \in b\urcorner$;
$\mathcal{I}$ of $\ulcorner otp(\mathfrak{X}) = \alpha\urcorner$ is $\ulcorner\mathfrak{X} \simeq mtk(\mathfrak{A})\urcorner$.

The last item reminds us that the von Neumann implementation of ordinals is available only in theories in which we can prove Mostowski's collapse lemma. In theories like Zermelo set theory we have to make do with Scott's trick.

Supremum and Infimum of sets of von Neumann ordinals are easily defined. $Sup(X) =: \bigcup X$ and $inf(X) =: \bigcap X$. The only grubby parts of the Von Neumann implementation are multiplication and addition, which both have to be defined by recursion. $\alpha+1 =: \alpha\cup\{\alpha\}$; $\alpha+\lambda =: \bigcup\{\alpha+\beta : \beta < \lambda\}$ for λ a limit ordinal. Similarly $\alpha\times(\beta+1) =: \alpha\times\beta+\alpha$; $\alpha\times\lambda =: \bigcup\{\alpha\times\beta : \beta < \lambda\}$ for λ a limit ordinal. $\alpha^{\beta+1} =: \alpha^{\beta}\times\alpha$; $\alpha^{\lambda} =: \bigcup\{\alpha^{\beta} : \beta < \lambda\}$ for λ limit. It is a nontrivial exercise to establish that these operations in fact implement the operations on ordinals arising from concatenation, cartesian product and functions-with-finite-support, but there are standard treatments in any elementary book on set theory. Finally in the von Neumann implementation T is clearly the identity.

6.4.2.1 *The Burali-Forti paradox for Von Neumann ordinals*

We shall characterise the class of Von Neumann ordinals as the class of all hereditarily transitive sets. But we preface this with a discussion of H_ϕ (the class of things that are hereditarily ϕ) for general ϕ.

A set is hereditarily ϕ if it is ϕ and all its members are hereditarily ϕ. Let H_ϕ be the collection of wellfounded sets that are hereditarily ϕ, the least fixed point for $\mathcal{P}_\phi()$ where $\mathcal{P}_\phi(x)$ is $\{y \in x : \phi(y)\}$. Thus H_ϕ is $\bigcap\{x : \mathcal{P}_\phi(x) \subseteq x\}$,

Remark 6.5 $\neg\phi(H_\phi)$

Suppose $\phi(H_\phi)$. Then $\phi(H_\phi) \in \phi(H_\phi)$. Then $\mathcal{P}_\phi(H_\phi) \setminus \{H_\phi\} \subseteq -\{H_\phi\}$ so $H_\phi \subseteq -\{H_\phi\}$ which is to say $H_\phi \not\in H_\phi$.

Next let x be arbitrary so that $\mathcal{P}_\phi(x) \subseteq x$. We have $H_\phi \subseteq x$ and $\phi(H_\phi)$ tells us that $H_\phi \in \mathcal{P}_\phi(x)$ whence $H_\phi \in x$ and—since x was arbitrary among x satisfying $\mathcal{P}_\phi(x) \subseteq x$—$H_\phi \in H_\phi$. This is a contradiction, so we must drop the assumption that $\phi(H_\phi)$. ■

Remark 6.6 *H_ϕ is transitive.*

Proof:
Let x be an arbitrary member of a member of H_ϕ. Thus for some y, and any X such that $\mathcal{P}_\phi(X) \subseteq X$, we have $x \in y \in X$. But $A \subseteq B \to \mathcal{P}_\phi(A) \subseteq \mathcal{P}_\phi(B)$ so $\mathcal{P}_\phi(\mathcal{P}_\phi(X)) \subseteq \mathcal{P}_\phi(X)$. Therefore $y \in \mathcal{P}_\phi(\mathcal{P}_\phi(X))$, and $\mathcal{P}_\phi(\mathcal{P}_\phi(X)) \subseteq \mathcal{P}_\phi(X)$ so $y \in \mathcal{P}_\phi(X)$, whence $y \subseteq X$ and $x \in X$. But X was an arbitrary X such that $\mathcal{P}_\phi(X) \subseteq X$, so x belongs to all such X and $x \in H_\phi$. But x was arbitrary. ■

It is now immediate that the collection H_{trans} cannot be a set. If it were a set it would be transitive by remark 6.6 but cannot be transitive, by remark 6.5. So we have a set-theoretic paradox: the paradox of the set of hereditarily transitive sets.

Next we check that H_{trans} is precisely the class of Von Neumann ordinals.

Remark 6.7 *$VN = H_{trans}$, the collection of hereditarily transitive (wellfounded) sets.*

Proof: $\in$ and $<$ agree on Von Neumann ordinals, so if $\gamma \in \beta \in \alpha$ we have $\gamma < \beta < \alpha$ then $\gamma < \alpha$ whence $\gamma \in \alpha$, so α is a transitive set. So every Von Neumann ordinal is transitive. Clearly all members of a Von Neumann ordinal are Von Neumann ordinals, (and therefore transitive) so Von Neumann ordinals are hereditarily transitive.

For the other direction we reason by *reductio ad absurdum* and let X be an $\in$-minimal hereditarily transitive set that is not a Von Neumann ordinal.

Let χ be the set of Von Neumann ordinals that are members of x. $X \neq \chi$ because X is not a Von Neumann ordinal. So there is $Y \in X$ which is not a von Neumann ordinal. But this Y is hereditarily transitive, contradicting $\in$-minimality of X. ■

Thus under the Von Neumann implementation of ordinals, the pun that is the Burali-Forti paradox is implemented as the set theoretic paradox of the class of hereditarily transitive sets. This paradox doesn't have a proper name, which is a pity since in any cirumstances in which there are two things that need to be distinguished it is helpful to have names for both of them rather than merely one.

It may be worth noting parenthetically that least fixed points are always likely to be paradoxical. Forster [2001] contains a proof of the following:

Remark 6.8 *Suppose f is monotone and injective:* $(\forall xy)(x \subseteq y \longleftrightarrow f(x) \subseteq f(y))$*. Let* $A := \bigcap\{x : \mathcal{P}(f(x)) \subseteq x\}$*. Then A is not a set.*

Proof: (omitted) ■

If we take f to be the identity we get Mirimanoff's paradox.

Notice that VN is paradoxical in a way that the collection of Russell-Whitehead ordinals is not. That collection does of course give rise to paradox in naïve set theory, but only because it enables one to prove the existence of V and thence the Russell class. As we have just seen, the paradoxicality of VN does not depend on this, but arises separately.

Bibliography

1971 Barwise, K.J "A Preservation theorem for interpretations" Lecture Notes in Mathematics 337, Springer-Verlag, New York, 1971, pp 618-21

1975 Barwise, K.J. Axioms for Abstract model Theory. Annals of Mathematical Logic 1975 pp 221-265

2001 Bechtel W, Mandik P, and Mundale J, "Philosophy meets the Neurosciences" in "Philosophy and the Neurosciences" ed W.Bechtel, P. Mandik, J. Mundale and R.S. Stufflebeam (eds) Blackwell 2001

1964 Benacarraf P. What numbers could not be. *in* Benacerraf and Putnam: Readings in the philosophy of mathematics. Basil Blackwell Oxford 1964

1897 Burali-Forti C. *Una questione sulli numeri transfiniti* **Rendiconti del Circolo Matematico di Palermo 11** 1897 pp 154-164; 260. *translated as: A question on transfinite numbers* and *On wellordered classes* in *From Frege to Gödel—A source book in mathematical Logic* ed. J van Heijenoort, Harvard University Press, pp 104-112

1947 Carnap, R. Meaning and Necessity. University of Chicago Press 1947

2001 Cleo Condoravdi, Dick Crouch, John Everett, Valeria Paiva, Reinhard Stolle, Danny Bobrow, Martin van den Berg. Preventing existence Proceedings of the international conference on Formal Ontology in Information Systems, eds N. Guarino and C. Welty, ACM Press, 2001. pp. 162 - 173 available on-line: www.parc.com/istl/members/crouch

1964 Coret, J. [1964] Formules stratifiées et axiome de fondation. *Comptes Rendus hebdomadaires des séances de l'Académie des Sciences de Paris série A* **264** pp. 809–12 and 837–9.

1979 Dennett, D. Brainstorms: Philosophical essays on mind and Psychology: Harvester Press, Hassocks 1979

1972 Enderton, H A mathematical introduction to Logic Harcourt/Academic Press NY and London 1972

1991 H. Eveking, 'Formal Verification of Synchronous Systems', in Formal Aspects of VLSI Design, edited by G. J. Milne and P. A. Subrahmanyam (North-Holland, 1991), pp. 137-151.

1993 Forster, T.E. The well-typed formulæ of λ-calculus are those preserved by a generalisation of Rieger-Bernays permutations. Theoretical Computer Science vol. **110**, (1993) 405-418

2001 Forster, T.E. Games played on an ill-founded membership relation, in the Boffa festschrift. Bull Maths Soc Belg, 2001

2003 Forster, T.E. ZF + "Every set is the same size as a wellfounded set" JSL to appear.

1903 Frege, G. Grundgesetze der Arithmetik ('Basic Laws of Arithmetic'), Jena: Verlag Hermann Pohle, Band I (1893), Band II (1903)

1967 Gauntt, R.J. Undefinability of cardinality, in Lecture Notes prepared in Connection with the Summer Institute on Axiomatic Set Theory, AMS 1967

1915 Hartogs. F. Über das Problem der Wohlordnung. Mathematische Annalen **76** pp 438-443.

1993 Hodges, W. Model theory *Encyclopædia of Mathematics and it applications* **42** CUP 1993

1999 Holtz, M. Stefens,K and Weitz, E. Introduction to cardinal arithmetic. Birhäuser: Basel, Berlin, Boston.

1987 Johnstone, P.T., Notes on Logic and set theory. Cambridge University Press, Cambridge, 1987

1960 Keisler H.J., and Tarski A., From Accessible to Inaccessible cardinals, *Fundamenta Mathematica* 1960.

1978 Kim J. supervenience and Nomological incommensurables. American Philosophical Quarterly 15,2 April 1978

1971 Krantz, D.H. *et al.* . Foundations of Measurement. Vol. 1. New York: Academic Press, 1971.

1975 Kripke Outline of a theory of truth. The Journal of Philosophy **72**, 690-716.

1969 Levy, A. The definability of cardinal numbers, Foundations of Mathematics, Gödel Festschrift, Springer-Verlag 1969, pages 15-38.

1983 Link, G. The logical analysis of plurals and mass terms: a lattice-theoretical approach, in Bäuerle, Schwarze and Stechow eds 'Meaning, Use, and Interpretation of Language' de Gruyter, Berlin, 302–323

2001 Mathias, A. R. D. The strength of Mac Lane set theory Annals of Pure and Aplied Logic **110** (2001) 107–233

1993 Melham, T.F., Higher Order Logic and Hardware Verification, Cambridge Tracts in Theoretical Computer Science 31 (Cambridge University Press, 1993).

1917 Mirimanoff, D. Les Antinomies de Russell et de Burali-Forti et le problème fondamental de la théorie des ensembles *L'Enseignement Mathématique* **19** (1917) pp 37-52.

1982 Gregory H Moore. Zermelo's axiom of choice: its origins, development and influence. Studies in the History of Mathematics and Physical Sciences v 8. Springer 1982

1966a Quine W. V. O. From a Logical Point of View in The Ways of paradox, Random House 1966

1966b Quine W. V. O. Three grades of modal involvement in The Ways of paradox, Random House 1966

1913 Russell B.A.W., and Whitehead A.N, Principia Mathematica, Cambridge University Press, Cambridge, (1910-1913).

1908 Russell, B. A. W. Mathematical logic as based on the theory of types. *American Journal of Mathematics* **30** pp. 222–62.

1903 Russell, B. A. W. The Principles of Mathematics. W.W. Norton New York.

1905 Russell, B. A. W. On Denoting MIND 1905

1919 Russell, B. A. W. Introduction to Mathematical Philosophy. Routledge and Keegan Paul, London

1962 Scott, D. S. Quine's individuals. In *Logic, methodology and philosophy of science.* Ed. E. Nagel, pp. 111–5, Stanford University Press Stanford 1962.

1992 Shelah, S. Cardinal Arithmetic for sceptics. Bull Am Maths Soc **26** no 2 (1992) pp 197-210

1950 Strawson. P.F., On referring MIND **59** pp 320-44.

1957 Suppes, P. Introduction to Logic Van Nostrand 1957. Also Dover 1999.

1959 Suppes, P. (1959) Measurement, empirical meaningfulness and three-valued logic. In C. W. Churchman and P. Ratoosh (Eds.), Measurement: Definitions and theories. New York: Wiley, 129-143.

1956 Tarski. Ordinal Algebras North-Holland (Studies in Logic and the foundations of mathematics) 1956

1973 Truss, J.K. Convex sets of cardinals Proc LMS **27** (1973) pp 577-99.

1983 Crispin Wright. Frege's conception of numbers as objects. *Scots Philosophical Monographs* **2** Aberdeen University Press 1983

Index of Definitions

Index